I0569224

Human Factors in Green Building

Human Factors in Green Building

Special Issue Editor
Zhonghua Gou

MDPI • Basel • Beijing • Wuhan • Barcelona • Belgrade

Special Issue Editor
Zhonghua Gou
Griffith University
Australia

Editorial Office
MDPI
St. Alban-Anlage 66
4052 Basel, Switzerland

This is a reprint of articles from the Special Issue published online in the open access journal *Buildings* (ISSN 2075-5309) from 2017 to 2019 (available at: https://www.mdpi.com/journal/buildings/special_issues/human_factors)

For citation purposes, cite each article independently as indicated on the article page online and as indicated below:

LastName, A.A.; LastName, B.B.; LastName, C.C. Article Title. *Journal Name* **Year**, *Article Number*, Page Range.

ISBN 978-3-03897-566-3 (Pbk)
ISBN 978-3-03897-567-0 (PDF)

Cover image courtesy of Zhonghua Gou.

ⓒ 2019 by the authors. Articles in this book are Open Access and distributed under the Creative Commons Attribution (CC BY) license, which allows users to download, copy and build upon published articles, as long as the author and publisher are properly credited, which ensures maximum dissemination and a wider impact of our publications.

The book as a whole is distributed by MDPI under the terms and conditions of the Creative Commons license CC BY-NC-ND.

Contents

About the Special Issue Editor

Zhonghua Gou is a Senior Lecturer in Architecture at Griffith University, Australia. His research interest in sustainable design and green building. He has published more than 60 peer reviewed journal articles in this field. He is also a regular journal reviewer and grants assessor. He has been recognized as the National Research Leader in Architecture by 'The Australian' in 2018. He has also been awarded with Griffith Learning and Teaching Scholarships for his approaches to incorporating environmental sustainability into architectural curricula.

Preface to "Human Factors in Green Building"

The idea about "Human Factors in Green Building" is derived from my PhD project "Addressing Human Factors in Green Building", in which I surveyed first generation green buildings in China using BUS (Building Use Studies) methodology. The aim of this project is to investigate or interrogate green design from a users' perspective. Numerous post-occupancy studies having been conducted across the world in recent decades. When I received the invitation from the Journal of Buildings to edit a Special Issue on green building, the idea came to my mind immediately, and I planned to collect papers focusing on human factors (such as occupant comfort, health, and wellbeing). The collected papers surprised me, and the scope of this Special Issue started to extend beyond my expectations. Targeted buildings and surveyed users are greatly diverse. This diversity and surprise further motivated me to work harder to publish my book. Of course, this book would have been impossible without the help of MDPI, especially the editorial team of Buildings. I want to thank Professor Derek Clements Croome; although we have never met, I can perceive his passion and inclination for helping young academics in the field of sustainable design.

I have to admit that this was my first time being a journal guest editor and I did not have much confidence in the beginning. Luckily, I came across many kind people while collecting papers. First of all, the research team from University of Melbourne, A/Professor Lu Aye, Dr. Hing-Wah Chau, etc., helped me tremendously, including through their papers and suggestions. Professor Stephen Lau and his team at National University of Singapore, Professor Baharuddin and his team at Hasanuddin University, and Professor Bin Cheng and his team at Southwest University of Science and Technology, all helped me tremendously too.

The Special Issue is not just a continuous collaboration with my old friends and colleagues in Asia, but it also allows me to get to know new friends and collaborators in this field. I went to the ASA (Architectural Science Association) conference 2017, held at Victoria University of Wellington, to present my research while calling for papers for this Special Issue. The conference provided me with great opportunities to meet many researchers in this field. During the conference, I invited several important contributions from Australia and New Zealand. Their contributions greatly diversified this Special Issue and this book. I did not go to Europe in recent years for research activities, although I am keen to do so. The contributions from UK and Sweden were a wonderful surprise and a gift.

Zhonghua Gou
Special Issue Editor

Editorial

Human Factors in Green Building: Building Types and Users' Needs

Zhonghua Gou

School of Engineering and Built Environment, Griffith University, Gold Coast, QLD 4215, Australia; z.gou@griffith.edu.au

Received: 17 December 2018; Accepted: 5 January 2019; Published: 9 January 2019

Abstract: The Special Issue on "Human Factors in Green Building" addresses the design of indoor environment quality for users' needs. The collected papers cover various building types and the research highlights the different needs of users. In working environments, employees' stress is the main concern in the workplace design, especially for open plan offices where lack of privacy and over exposure to environmental stress have been reported. In residential environments, residents have great opportunities to adjust their environments to suit their needs; therefore, passive design such as natural ventilation is explored in residential buildings with climates such as cold or humid tropical. In healthcare environments, the papers in this issue are concerned with the needs of patients, especially the older adults who require special care. In learning environments, thermal and visual aspects are investigated for optimal comfort conditions and learning outcomes. The special issue demonstrates insightful critical thinking of indoor environment quality and proposes a new understanding for more practical design solutions. This editorial note is a brief review of the 12 papers, concluding with reflections about design of built environments to meet users' needs.

Keywords: human factors; green building; indoor environment quality; building types; post-occupancy evaluation

1. Introduction

Buildings serve their users and users adapt to their buildings. The intricate relationship should be addressed in the sustainable or green built environment design [1]. The research challenge is how we understand and measure human factors. In recent years, numerous human factors related studies have been found on the subject of indoor environment quality (IEQ) which includes aspects such as thermal comfort, air quality, noise and visual aspects of a building [2–4]. These studies aimed to define optimal settings for satisfying building occupants. On the other hand, the studies on IEQ have been facing great challenges due to its narrow definition of a physical environment that influences occupants' perception and satisfaction [5] and also its ignorance of building types and related diverse occupants' needs [6]. The aim of this special issue is to enrich the understanding of IEQs in relation to building types and users' needs.

To meet this aim, the special issue collected 12 papers from a variety of perspectives in response to the human factors. The authors come from Australia, China, Indonesia, U.K., U.S., Sweden and New Zealand, and represent worldwide efforts on this topic. These papers provided innovative frameworks theoretically and empirically to measure human-related IEQs in different building types. One of most mature IEQ studies is thermal comfort, a regular topic in this field. This special issue collected papers on measuring thermal comfort in vernacular architecture made of stone in cold climates, thermal comfort in tropical classrooms and thermal comfort and related use behaviors in modern apartment buildings. The contrast makes the special issue interesting to read and compare. The issue also collected papers on visual aspects (outdoor views) for enhancing indoor environment quality in learning environments. Stress or negative perception in contemporary open plan workplace is an urgent issue for addressing human factors in green building. The special issue collected two papers that aim to propose design solutions in response to workplace stress. Healthcare facilities are intensively researched for occupant health and wellbeing. For such a regular topic in this field, the related design strategies are analyzed in three papers. Last but not least, this special issue invited a special contribution that extended the current understanding of IEQ. The research about human factors is in great variety and covers different types of buildings. Their users have different expectations and needs. To enhance the environmental quality and meet their needs is what the special issue wants to promote for human-oriented design solutions. The following editorial comments on the collected papers are based on the building types and users' needs.

2. Building Types and Users' Needs

2.1. Workplace Environments

In working environments, office workers are exposed to environmental and job stresses. Especially in open plan workplaces, the environmental stress due to loss of privacy and exposure to noise is outstanding. Three papers in this special issue offer design solutions to alleviating the workplace stress and improving indoor environmental conditions. The paper by Felix Kin Peng Hui and Lu Aye [7] established a comprehensive framework about health relevance of both proximal and remote aspects of workplace design. The method is a comprehensive literature review. It covered not only employees' immediate work area and ambient environmental qualities of the work area, but also building organisation, exterior amenities, and site-planning. The paper addressed that occupational stress is a complex phenomenon that is dynamic and evolving over time and developed an improved model relevant to work place design and occupational stress. The proposed improved model is presented with an appropriate causal loop diagram to assist in visualizing how different variables in a system are interrelated. The developed model highlights how connection to nature in workspaces can function as a work resource with a dual effect of improving physical wellbeing and psychological wellbeing.

The paper by Zhonghua Gou, Jian Zhang and Leigh Shutter [8] presents an empirical study about the individual environmental control, especially its benefits on self-reported health status in open-plan work environments where often occupants reported loss of privacy and negative feelings of their health status. The research combined three systematic occupant survey tools and collected responses on 12 selected individual controls. The results showed that half of the 12 individual controls were negatively associated with adverse perceptions. Among all controls, non-mechanical ones, such as windows and blinds, were more effective than mechanical ones such as fans and air-conditioning in alleviating adverse perceptions in open-plan offices. The research provides some interesting findings to workplace and interior design.

One paper in this special issue reported an experiment in a laboratory with workplace settings. It is different from other field studies of real offices mentioned above. It aims to investigate human factors in a controlled condition. This paper is contributed by Mattias Holmgren and Patrik Sörqvist [9]. Their experiment explored whether green building certification could make people favor that office environment over than the non-certified building. Through two rigorously deployed experiments, the author suggested the complexity of the green effect on influencing occupants' preference on indoor environmental conditions. The research had important implications for using green certification to improve office workers' satisfaction.

2.2. Healthcare Environments

Healthcare environmental design, especially for the elderly, have attracted intensive research attention in recent years. In this special issue, there are three papers relevant to this topic.

The paper by Hing-wah Chau, Clare Newton, Catherine Mei Min Woo, Nan Ma, Jiayi Wang and Lu Aye [10] focused on the specific group of people with dementia. This paper investigated three recently constructed dementia support facilities in Victoria, using fieldwork observation, design evaluation and space syntax analysis. The results provided evidence and critical analysis concerning the design of these three facilities on how the built environment can best accommodate residents with dementia.

Yisong Zhao and Monjur Mourshed [11] conducted an interesting study about the hospital outpatient area using a survey of patients. This study is in response to the increasing interest in 'patient-centred' design of health and care facilities, especially the hospital waiting areas. The research randomly selected outpatients in two hospitals in Qingdao, China for the survey. Five principal factors, respectively on sensory, lighting and thermal, facilities, spatial, and seating design, were identified. A variety of demographic data were correlated with these factors.

Another article on aged care facilities by Masa Noguchi, Nan Ma, Catherine Mei Min Woo, Hing-wah Chau and Jin Zhou [12] aimed to incorporate indoor environment quality into the architectural practice of aged cared facilities through the proposed environmental experience design framework. The method is case study. The authors extended their previous case study on the collective spatial analysis and IEQ monitoring results to apply the framework to the aged care facility in Victoria, Australia. This study helped to engage the subjectivity and objectivity of end users' expectations, desires, and requirements in the architectural design thinking process.

2.3. Learning Environments

Compared to working and residential environments, learning environments have attracted less research attention due to the group of people who may be under age to participate in the research. The special issue is lucky to have two important contributions, respectively concerning visual and thermal environments for learning activities. The paper by Zhonghua Gou, Maryam Khoshbakht and Behnam Mahdoudi [13] reported a case study of outdoor views in a university library in the south hemisphere. The study surveyed the students' seat preference in the selected library and addressed the importance of outdoor views in their seat selection. Furthermore, the study quantified the outdoor view in terms of sky view, tree view and shading view. The three views correlated with the seat preference to different extents.

In the other paper about learning environments by Baharuddin Hamzah, Zhonghua Gou, Rosady Mulyadi and Samsuddin Amin [14], the thermal comfort level of students in secondary schools in the tropical city of Makassar was measured and analyzed. It is a large-scale study with 1594 students in 48 classrooms under natural ventilation. It turned out that the air temperatures ranged from 28.2 °C in the morning to 33.6 °C in the midday, which is out of the normal comfort zone defined in many building standards. Nevertheless, the students did not report much discomfort; instead, most of them accepted the thermal conditions. Neutral temperatures were identified for the group of students in Indonesia. This paper echoed another paper [15] in this special issue which was concerned with residential environments in tropical climates and pointed out that the tropical people could adapt to temperatures higher than the normal accepted comfortable temperature.

2.4. Residential Environments

Different from institutional building types such as office and school buildings, residential buildings are designed to meet demographic needs of residents. Therefore, the research focus is usually in a great diversity. The special issue collected three papers in residential environments. The paper by Zhonghua Gou, Wajishani Gamage, Stephen Siu-Yu Lau and Sunnie Sing-Yeung Lau [15] is a pilot study of thermal comfort and adaptive behaviors of occupants who live in naturally ventilated dormitories at the campus of the National University of Singapore. The research used a longitudinal survey and field measurement to measure thermal comfort, adaptive behaviors and indoor environment qualities. Although occupants living in naturally ventilated buildings in tropical climates were exposed to higher operative temperatures than what comfort standards recommend for naturally conditioned spaces, they still felt that such conditions were acceptable. This finding echoes the adaptive thermal comfort theory proposed by de Dear and Brager [16]. The study also found two important behavioral adjustments that contributed to the acceptance: increasing the indoor air velocity and reducing clothing insulation.

The paper concerning residential environments in cold climates is contributed by Bin Cheng, Yangliu Fu, Maryam Khoshbakht, Libin Duan, Jian Zhang and Sara Rashidian [17]. Different from the other one that is focused on modern apartment buildings, this paper focused on traditional residential building made of stone. The research conducted thermal comfort measurements in winter. The majority of surveyed residents voted "slightly cool" for temperature, and "slightly dry" for humidity. The available adaptive opportunities for the residents included adjusting clothing, drinking hot beverages, blocking air infiltration through windows, and changing activities.

Yukiko Kuboshima, Jacqueline McIntosh and Geoff Thomas [18] used a qualitative case study approach to investigate the rental housing design for the elderly. The method consisted of a detailed documentation of the physical environment, followed by interviews with and full-day observations of the residents and their caregivers, to examine the living experiences of six old people who lived in local-authority rental housing in New Zealand. The authors found that the design of housing that improves their life quality requires solutions to accommodate the various conflicting needs derived from the diversity in the user's preferences and impairments. Particularly, there was greater need for additional or reorganized space to accommodate caregivers and visitors while to maintain residents' independence, privacy, and other aspects important for their life quality.

2.5. Theory Attempts to Fill the Gap

Most research on indoor environment quality is based on survey, field or laboratory experiments or case study; few explored its theories or related arguments. The paper contributed by Linda Pearce [19] fills in the gap. The paper used a theoretical method. The paper highlights the pleasure of interior environments. Specifically, the paper proposed a sequential mixed methods research process allowing subjective and objective research methods integration. The methods integrated interior architecture and architectural science disciplines by coding interior architecture perspectives into possible measurable variables which would likely be more inclusive of the lived experience and agency of occupants of interior spaces. The paper had important implications for expanding indoor environment quality indicators.

3. Concluding Remarks

The traditional understanding and measuring of human factors are based on quantitative studies of IEQs to identify the optimal range of indoor environmental settings that can satisfy users. However, the reality is that building users and their needs are different; the quantitative understanding of IEQs might be flawed. The special issue contains both traditional quantitative measurement of IEQs on thermal comfort and visual aspects, and newly proposed IEQ frameworks that contain spatial experience, facilities and even aspects superseding the physical boundary of a building. More importantly, these papers diversify the needs of IEQs according to building types: workplace, residential, learning and healthcare environments. These papers addressed important IEQ and design issues for the specific building types and users' needs. An important message from this special issue is that the future study of IEQ requires going beyond the discipline of building or architectural sciences to include and integrate other disciplines such as interior design, healthcare design, workplace design and environmental psychology.

Funding: This research received no external funding.

Acknowledgments: Finally, I would like to thank all anonymous reviewers for their constructive, critical comments, all the authors for responsive, responsible revisions, and of course the team at MDPI for their efficiency in the whole publishing process.

Conflicts of Interest: The author declares no conflict of interest.

References

1. Baird, G. *Sustainable Buildings in Practice: What the Users Think*; Routledge: London, UK, 2010.
2. Gou, Z.; Prasad, D.; Lau, S.S.-Y. Impacts of green certifications, ventilation and office types on occupant satisfaction with indoor environmental quality. *Arch. Sci. Rev.* **2014**, *57*, 196–206. [CrossRef]
3. Khoshbakht, M.; Gou, Z.; Xie, X.; He, B.; Darko, A. Green building occupant satisfaction: Evidence from the australian higher education sector. *Sustainability* **2018**, *10*, 2890. [CrossRef]
4. Piasecki, M.; Kozicki, M.; Firlag, S.; Goljan, A.; Kostyrko, K. The approach of including tvocs concentration in the indoor environmental quality model (ieq)—case studies of breeam certified office buildings. *Sustainability* **2018**, *10*, 3902. [CrossRef]
5. Gou, Z.; Prasad, D.; Siu-Yu Lau, S. Are green buildings more satisfactory and comfortable? *Habitat Int.* **2013**, *39*, 156–161. [CrossRef]
6. Xue, F.; Gou, Z.; Lau, S. Human factors in green office building design: The impact of workplace green features on health perceptions in high-rise high-density asian cities. *Sustainability* **2016**, *8*, 1095. [CrossRef]
7. Hui, F.; Aye, L. Occupational stress and workplace design. *Buildings* **2018**, *8*, 133. [CrossRef]
8. Gou, Z.; Zhang, J.; Shutter, L. The role of personal control in alleviating negative perceptions in the open-plan workplace. *Buildings* **2018**, *8*, 110. [CrossRef]
9. Holmgren, M.; Sörqvist, P. Are mental biases responsible for the perceived comfort advantage in "green" buildings? *Buildings* **2018**, *8*, 20. [CrossRef]
10. Chau, H.-w.; Newton, C.; Woo, C.; Ma, N.; Wang, J.; Aye, L. Design lessons from three australian dementia support facilities. *Buildings* **2018**, *8*, 67. [CrossRef]
11. Zhao, Y.; Mourshed, M. Patients' perspectives on the design of hospital outpatient areas. *Buildings* **2017**, *7*, 117. [CrossRef]
12. Noguchi, M.; Ma, N.; Woo, C.; Chau, H.-W.; Zhou, J. The usability study of a proposed environmental experience design framework for active ageing. *Buildings* **2018**, *8*, 167. [CrossRef]
13. Gou, Z.; Khoshbakht, M.; Mahdoudi, B. The impact of outdoor views on students' seat preference in learning environments. *Buildings* **2018**, *8*, 96. [CrossRef]
14. Hamzah, B.; Gou, Z.; Mulyadi, R.; Amin, S. Thermal comfort analyses of secondary school students in the tropics. *Buildings* **2018**, *8*, 56. [CrossRef]
15. Gou, Z.; Gamage, W.; Lau, S.; Lau, S. An investigation of thermal comfort and adaptive behaviors in naturally ventilated residential buildings in tropical climates: A pilot study. *Buildings* **2018**, *8*, 5. [CrossRef]
16. De Dear, R.; Schiller Brager, G. The adaptive model of thermal comfort and energy conservation in the built environment. *Int. J. Biometeorol.* **2001**, *45*, 100–108. [CrossRef] [PubMed]
17. Cheng, B.; Fu, Y.; Khoshbakht, M.; Duan, L.; Zhang, J.; Rashidian, S. Characteristics of thermal comfort conditions in cold rural areas of china: A case study of stone dwellings in a tibetan village. *Buildings* **2018**, *8*, 49. [CrossRef]
18. Kuboshima, Y.; McIntosh, J.; Thomas, G. The design of local-authority rental housing for the elderly that improves their quality of life. *Buildings* **2018**, *8*, 71. [CrossRef]
19. Pearce, L. Translating across disciplines: On coding interior architecture theory to advance complex indoor environment quality. *Buildings* **2018**, *8*, 82. [CrossRef]

© 2019 by the author. Licensee MDPI, Basel, Switzerland. This article is an open access article distributed under the terms and conditions of the Creative Commons Attribution (CC BY) license (http://creativecommons.org/licenses/by/4.0/).

Article
Occupational Stress and Workplace Design

Felix Kin Peng Hui and Lu Aye *

Renewable Energy and Energy Efficiency Group, Department of Infrastructure Engineering,
Melbourne School of Engineering, The University of Melbourne, Melbourne, Vic 3010, Australia;
huik1@unimelb.edu.au
* Correspondence: lua@unimelb.edu.au; Tel.: +61 3 8344 6879

Received: 16 August 2018; Accepted: 21 September 2018; Published: 23 September 2018

Abstract: The World Green Building Council (WGBC) advocates improvements in employee health, wellbeing, and productivity in buildings as people are about 90% of an organisation's expense and well exceed building costs and energy costs. It was reported that earlier research on workplace design primarily focused on physical arrangement of employees' immediate work area, and ambient environmental qualities of the work area. Building organisation, exterior amenities, and site-planning have been given less attention. Therefore, we examine more closely the health relevance of both proximal and remote aspects of workplace design. Occupational stress is a complex phenomenon that is dynamic and evolving over time. This investigation reviews the existing fundamental conceptual models of occupational stress, workplace design, and connection to nature. It aims to develop an improved model relevant to work place design and occupational stress linked with connection to nature. The proposed improved model is presented with an appropriate causal loop diagram to assist in visualizing how different variables in a system are interrelated. The developed model highlights how connection to nature in workspaces can function as a work resource with a dual effect of improving physical wellbeing and psychological wellbeing.

Keywords: occupational stress; workplace design; connection to nature; wellbeing; causal loop diagram

---

1. Introduction

The World Green Building Council (WGBC) provides a compelling business case argument for health, wellbeing, and productivity: "people are 90% of an organisation's expense and well exceed building costs and energy costs, therefore, a small improvement in employee productivity can yield significant value." [1] (p. 2). The relationships between workplace design and occupational health were examined by Stokols (2011) [2], who reported that they are considered at several levels of analysis:

1. Physical arrangement of employees' immediate work area,
2. Ambient environmental qualities of the work area,
3. Physical organization of buildings that comprise a particular workplace,
4. Exterior amenities and site planning of those facilities.

Stokols (2011) [2] found that workplace design levels 1 and 2 were the primary areas of focus of earlier research, while level 3: building organisation and level 4: exterior amenities and site-planning have been given less attention. Stokols (2011) [2] recommended a thorough examination of the health relevance of both proximal and remote aspects of workplace design in future research. In this paper, we examine more closely the connection to nature and wellbeing aspects of workplace design.

Based on survey data, Higginbottom (2014) [3] reported that employees suffering from high stress levels have lower engagement, are less productive, and have higher absenteeism levels than those not working under excessive pressure. Occupational stress is a complex phenomenon that is dynamic and evolving over time. The traditional study designs for occupational stress have been statistical tests.

To better capture the underlying dynamic processes including feedbacks and delays, system dynamics (an analytic method) is more appropriate. This investigation reviews the existing conceptual and system dynamics models of occupational stress. None of them include the connection to nature aspect, even though positive health effects of this aspect have been documented in biophilic design literature. It aims to develop an improved model relevant to work place design and occupational stress linked with connection to nature. The proposed improved model is presented with an appropriate causal loop diagram to assist in visualizing how different variables in a system are interrelated.

The World Health Organization (2007) [4] (p. 4) defined work-related stress as "a pattern of reactions that occurs when workers are presented with work demands not matched to their knowledge, skills, or abilities and that challenge their ability to cope". This is similar [5] to the most cited definitions of occupational stress by the National Institute for Occupational Safety and Health (NIOSH) [6]. One of the conceptual models of stress, the Job Demands–Resources (JD–R) model [7], presented in Section 2, aligns well with these definitions. Hobfoll (2002) [8] categorised resources into four types: status resources, material resources, social resources, and personal resources. The Conservation of Resources (COR) theory by Hobfoll (1989) [9] emphasized resources that may be categorised into personal (metal, physical, and social) and job-related. In this investigation, we apply the COR theory [9] and combine it with the JD–R model [7] to explain the mechanism by which a job resource of having a work environment, which enable connection to nature, can reduce stress, which in turn alleviates the negative effect of a high job demand situation.

In particular, the combined JD–R model and COR model is used to show that having an environment, which enable connection to nature, has a positive effect on organizational outcomes.

The reasons are as follows:

1. Physical resource that can directly promote wellbeing. Some may argue that green natural environments are a physical resource.
2. No direct relation to condition or personal characteristics.
3. Energies—seen as value in aiding the other type of resources. Especially, it is seen as an aid to maximizing the use of relevant personal traits in that it promotes concentration and a sense of awareness.

The main aim of this review is to propose an improved conceptual model that incorporates three themes: wellbeing, workplace design, and connection to nature. Two innovative aspects of the improved model are the following: (1) it includes nature as a resource or a pre-requisite in workplace design; (2) it explains the mechanism of how these three themes are connected.

2. Method

The existing fundamental conceptual models of occupational stress (Job Demands–Control (JD–C) model, Job Demands–Resources (JD–R) model, and Conservation of Resources (COR) model), and workplace design, connection to nature, and biophilic design were reviewed. Literature searches were conducted using Web of Science [10] for the time span between 1900 and 2018 on the core database. It should be noted that the review is based on literature published in English only. The numbers of documents found for various combinations of keywords are shown in Table 1.

Table 1. Web of Science search results.

No.	Search Terms	Result
1	"Job demands–control model"	48
2	"Job demands–resources model"	419
3	"Conservation of resources model"	33
4	"workplace design"	387
5	"connection to nature"	97
6	"biophilic"	162
7	"Job demands–control model" AND "connection to nature"	0
8	"Job demands–resources model" AND "connection to nature"	0
9	"Conservation of resources model" AND "connection to nature"	0
10	"workplace design" AND "connection to nature"	0

From the results of the literature search, a mix of highly cited papers and recent papers were selected for review with a focus on the models' strengths and limitations. The next section presents the conceptual models found in the literature. Based on the findings of the review, we proposed a modified occupational stress model that considers connection to nature as a resource in Section 5.

3. Conceptual Models on Occupational Stress

Research over the years has shown that job characteristics influence an employee's wellbeing [7,11–14]. High work pressure and job demands may lead to negative health outcomes such as high blood pressure, headaches, indigestion, and insomnia [15]. On the other hand, the provision of adequate job resources can help employees with a high demand job environment and can help alleviate the risk imposed by the demands of any job [16,17]. The nature of job resources is also a topic that needs detailed examination as these are not restricted to characteristics of the work environment and may include personal resources such as resilience, ability to cope with stress, and stress coping behaviours [9]. Recent research also showed that personal characteristics influence organizational outcomes [13,14,18]. In the following sections, we present some of the early classic work design research on job demands and job control models, job demand and resources model, and the inclusion of personal characteristics in these models. This early research has been developing over the years and lays the important foundation framework to explain the relationship of work design to employee wellbeing. The importance of personal characteristics in dealing with the job demands are also important in determining the level of contribution to organizational outcomes [18].

3.1. Job Demands-Control (JD-C) Model in Classic Work Design Theory

Karasek's Job Demands–Control (JD–C) model [11] is the earliest and most cited model that relates work design to occupational stress [19]. In the Job Demands–Control model, the ability to control a piece of work would alleviate the negative effects of job demands. It also helps enhance employees' job satisfaction and engages them in more challenging tasks and jobs requiring higher-level skillsets [11]. In the JD–C model (Figure 1), Karasek [11] postulated that in regards to labour intensity, the work environment has an effect on health promotion (reduces work strain). On one hand, the level of work strain or job demands includes requirements such as work rate, time allocated, the anticipated pressure to complete a task, effort, and relative difficulty. Such requirements contribute to the psychological stressors in the work environment. On the other hand, the level of control afforded by the workplace, also known as job decision latitude, determines the freedom that an employee has in initiating, organizing, executing, and controlling his own work. This job decision latitude refers to the control that employees have in going about their duties and the manner in which the tasks are performed [20]. It concerns both the employee's internal resource of competence and an external factor such as the decision-making authority.

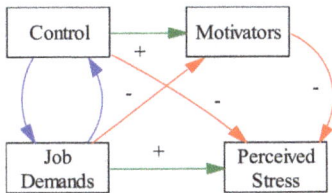

Figure 1. Job Demands–Control (JD–C) model (adapted from Wallgren and Hanse, 2007 [21]). Note: blue arrows indicate co-varying determinants, green arrows indicate positive causal relations, and red arrows indicate negative causal relations.

One important managerial requirement of the JD–C model is the need to have a balance for job autonomy and work stress, as Karasek [11] postulated that employees in high demand jobs would experience high stress if they cannot decide how to manage the work. The element of control would cause employees far less stress than if they had no control at all. This is especially true when employees have to work within given time schedule or budget constraints. High work pressure coupled with low amount of control would increase the risk of stress. Although the JD–C is simplistic in that it attempts to link psychological fatigue or work stress to two elements of job demands, namely level of autonomy and control, it lays the groundwork to offer insights into the types of managerial interventions that may be possible.

Some researchers cited inconsistencies in the results obtained with the Job Demands–Control model [19,22]. One common explanation was that different variables have been used to measure demands, control, and strain, and more importantly, they do not take workers' personal characteristics into account [22].

3.2. The Job Demands–Resources (JD–R) Model

Research has shown that job characteristics influence an employee's wellbeing. High work pressure and job demands may lead to health problems such as high blood pressure and insomnia. On the other hand, the provision of adequate job resources can help employees in a high demand job environment and can help alleviate the risk imposed by the demands of any job. Over the years, researchers made attempts to improve on the JD–C, adding to the model with useful constructs such as resources, burnout, self-efficacy, and social support [12,23–25]. The Job Demands–Resources (JD–R) model is recognized as one of the leading job stress models. A reason for this, as explained by Schaufeli and Taris (2014) [26], is that it is more flexible and can be tailored to a much wider variety of work settings.

The JD–R model [12] focused on the effects of resources–demand on employee burnout using Maslach's burnout inventory of exhaustion, cynicism, and self-efficacy. Later research moved the focus towards employee disengagement and engagement and this had wider managerial implications for organizational effectiveness. Potentially, it includes all job settings and job environments. The JD–R model presented by Bakker and Demerouti in 2007 [7] explains how these two concepts of job demands and job resources interact to produce strain and motivation, respectively, in determining organizational outcomes (Figure 2). Job demands refer to those physical, psychological, social, or organizational requirements of the job and job resources refer to those physical, psychological, social, or organizational enablers of the job. Job demands can lead to strain, which in turn is negatively correlated with organisation outcomes. Job resources are negatively correlated with job demands and positively correlated with motivation and organizational outcomes.

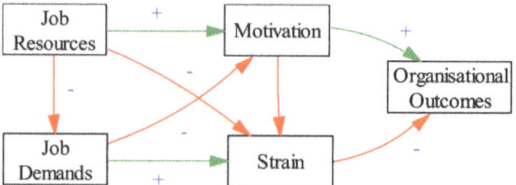

Figure 2. The Job Demand–Resources (JD–R) model (adapted from Bakker and Demerouti, 2007 [7]). Note: green arrows indicate positive causal relations, and red arrows indicate negative causal relations.

The JD–R model theorises that organizational outcomes are dependent on the motivation and mental wellbeing (here represented as strain), which in turn stems from a balance of positive aspects (job resources) and negative aspects (job demands) of job characteristics. Researchers have also applied these to a variety of settings including longitudinal settings, varying sample sizes, and different types of work environments. In a study of 342 hospitality workers working in 120 work units comprising 60 hotel front desks and 60 restaurants, Salanova, Agut, and Peiro (2005) [27] examined the service climate and employee work engagement and found these to be linked to employee performance and customer loyalty. In 2013, Hui [18] investigated 230 workers in the manufacturing sector in Australia and linked employee engagement, appropriate goal orientation, and the need for achievement to innovative behaviour at work. The presented results [18] suggest that providing work units with organizational resources increased collective work engagement, which in turn helped foster an excellent service climate and performance. A Finnish study of dentists in 2014, by Hakanen and Koivumaki [28], found work engagement was positively associated with the amount of procedure fees and consequently with dentists' pay level. However, exhaustion was not correlated with productivity. In a longitudinal study of 274 Spanish secondary school teachers, Llorens-Gumbau and Salanova-Soria (2014) [29] found the difficulty of the task (job requirements) is positively related to burnout, which in turn is negatively related to self-efficacy. Likewise, enablers such as resources are positively related to engagement and self-efficacy. As suggested in the JD–R model, good organizational outcome is a result of the mediating effects of good mental health on job demands and the mediating effects of motivation on job resources.

3.3. Stress and the Conservation of Resources (COR) Model

While the relationships between job resources and strain in the JD–R model are clear, Hofoll (1989) [9] presented an explanation for conceptualizing stress (or work strain) using a resources concept. While recognising that there are different views of stress such as the physical view of stress, a stimulus view, or an event perspective of stress, and that there are problems associated with these, Hofoll (1989) [9] suggested that stress results from the net gain or loss of resources and that people have four types of resources they work with in their everyday lives. This model (Figure 3) explains behaviour that people retain, protect, and build resources. Psychological stress would result from (a) a threat or a net loss of resources, (b) the net loss of resources, and (c) net loss after or a lack of gain after investment of resources [30]. Hofoll (1989) [9] said that perceived and actual losses are both valid sources of strain.

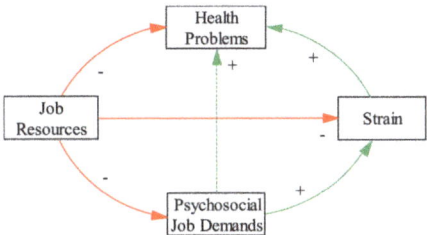

Figure 3. The Conservation of Resources (COR) model (adapted from Mayerl et al., 2016 [14]). Note: green arrows indicate positive causal relations, and red arrows indicate negative causal relations.

According to the COR model, there are four types of resources [9]. An object (physical) resource has some useful physical characteristics that would directly affect the ability of a person to work. It also has value because it indicates status such as a home or a luxury car. The second type of resource is a condition such as marital status, tenure, seniority, graduation, and so on. This type of resource is sought after because it provides promise for future jobs. The third type of resource is personal characteristics such as traits, skills, and abilities. This directly affects a person's ability to resist stress [31]. The last type of resource is energy, which refers to the intrinsic values of a person such as time or support, for example, having a large social network. Wellman (1981) [32] found that having a large social network contributed to the energies of a person. The COR model suggests that people react to stressful situations (loss or potential loss of resources) by employing other resources to offset the loss. Symbolic replacement may also replace direct replacement. The model suggests that people are motivated by the net gain of resources and may put up with temporary stressful states to gain resources. The COR model is useful in that it offers an insight into resources and strain and what motivates people to use resources to overcome strain.

3.4. Incorporating the COR in the JD–R Model

The JD–R model considered that the job resources were very much a part of the work environment. However, researchers such as Mayerl et al. (2016) [14] have argued that human behaviour comes as a result of people's interaction with the physical environment. Personal resources are normally associated with qualities such as resilience and usually enable a person to control his physical environment. Mayerl et al. (2016) [14] found that personal resources can be used in directing activities to improve wellbeing, as well as influence the perception of job characteristics. They posit that personal characteristics should not exist separately and should be part of the resources available to an employee.

Mayerl et al. (2016) [14] considered that the JD–R model overemphasized the environment external to the individual and neglected the individual and personal characteristics such as energy level, time, or resilience. Making use of Hobfoll's (2003) [33] COR model, which defines qualities such as resilience and optimism into four types of personal resources, Mayerl et al. (2016) [14] went on to develop an integrated model that shows how personal resources can attenuate negative effects of high job demands. The combined model recognises Hobfoll's (1989) COR theory [9], in how people strive to preserve their resources and resources do not exist in isolation, as well as in how people build resources to generate other resources. Individuals high in personal resources are also able to create resources, resulting in a high resource environment [13,34]. Mayerl et al.'s (2016) [14] model combines both job and personal resources to a common resources factor and linked this factor to the health-impairment process of the JD–R model. Mayerl and his team [14] surveyed 8657 participants from the Austrian working population and found that job and personal resources can be considered as indicators of a single resources factor, which was negatively related to psychosocial job demands, mental strain, and health problems. Confirming previous studies, they further found that mental strain mediated the relationship between psychosocial job demands and health problems. Their findings suggest that interventions aimed at maintaining health in the context of work may act on three levels: (1) the prevention of extensive job demands, (2) the reduction of work-related mental strain, and (3) the strengthening of resources.

3.5. The Importance of Creating Positive Emotions at Work

Work engagement is defined as a positive, fulfilling, work-related state of mind that is characterized by vigour, dedication, and absorption. Engaged workers are full of energy (vigour), strongly believe in their work (dedication), and are often fully concentrated and happily engrossed in their work activities in a sense of flow in which time passes quickly (absorption). Job resources have been shown to enhance work engagement (Xanthopoulou et al., 2007) [13]. Work engagement in turn has positive effects on organisational outcomes. In a worldwide study involving 7939 business units in 36 companies by the Gallup Organisation, Harter, Schmidt, and Hayes (2002) [35] reported that employee satisfaction and employee engagement has a considerable effect on business outcomes. Changes in management practices that enhance employee engagement will bring about positive business outcomes including profits.

Humans have a natural affinity for nature and biophilic design of workspaces helps enhance all three components of work engagement by bringing humans closer to nature (increase in vigour, dedication, and improved flow at work). Connection to nature in workspaces can be seen as bringing about an increase in internal personal resources. This give rise to positive emotions, which can bring about work engagement and positive effects on organisational outcomes.

The next section presents system dynamics models of occupational stress.

4. System Dynamics Models on Occupational Stress

System dynamics methodology enable understanding and describing counter-intuitive behaviour of complex systems. Two system dynamics models that quantify time varying occupational stress related parameters are available in the current literature. They were reviewed with a focus on work place design and connection to nature aspects in this section.

4.1. Morris et al. 2010 Model

Morris et al. (2010) [36] presented the definitions of following terms: Stress, Eustress, Distress, Demand, Resource, three Coping styles (Action-oriented, Emotion-oriented, Avoidant-oriented), three Degrees of stability (Emotional, Biological, Cognitive), Response action, two Locus of control (positive, negative), Cortisol level, Anxiety, and Heart rate as described in Hobfoll (1988) [37]. A simplified causal loop diagram (CLD) presenting the model is shown in Figure 4. The simulation time step used in Simgua's (2018) [38] model was one hour and the analysis period was 12 h. As shown in Figure 4 stress is the outcome resulted from demands, resources and job control. Perceived problem demands and stress are "positively correlated (i.e., when one increases or decreases so does the other)" [36] p. 4371. Perceived resources and stress are "negatively correlated (i.e., when one increases or decreases the other does the opposite)" [36] p. 4371. Job control and stress are also negatively correlated. Stress and health problem indicators (cortisol level, anxiety, and heart rate) are positively correlated. The model is able to simulate the behaviour of stress in a quantifiable manner for the 12 h analysis period. However, no work place design nor connection to nature aspects are considered by the model.

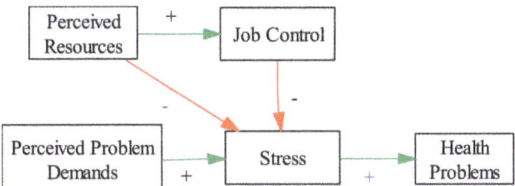

Figure 4. Morris et al. (2010) model [36] presented in a casual loop diagram. Note: green arrows indicate positive causal relations, and red arrows indicate negative causal relations.

4.2. Jetha et al. 2017 Model

A multidimensional system dynamics model (SDM) of workplace stress among nursing aides was developed by Jetha et al. (2017) [39]. By applying the model, they conducted simulations to illustrate how changes in psychosocial perceptions and workplace factors might influence workplace stress over time. Perceived workplace stress, Job demands, Job control, Job resources (Workplace social support) were considered (Figure 5). The analysis period used in their Vensim (Ventana Systems, 2018) [40] model was 10 weeks. Job demands and perceived workplace stress are positively correlated and job control and stress are negatively correlated as in Morris et al.'s (2010) model [36]. However, perceived workplace stress is considered as one of the determinants for job resources (note the backword arrow to job resources in Figure 5) and they are negatively correlated. The model is able to simulate and quantify the effects of changes in job control, job demands and job resources (workplace social support) on perceived workplace stress. However, no parameters directly related to work place design and connection to nature aspects are mentioned by the model.

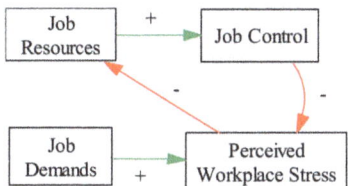

Figure 5. Jetha et al. (2017) model [39] presented in a casual loop diagram. Note: green arrows indicate positive causal relations, and red arrows indicate negative causal relations.

5. Green Buildings, Wellbeing and Connection to Nature

Green building rating tools have been developed to encourage and incentivise pushing the boundaries on sustainability. In this section how these tools address wellbeing and relate connection to nature have been explored. Work stress aspects of biophilic design, which aims to make use of natural elements in architectural and environmental design, is also investigated (Section 5.2). The relationships between connection to nature and wellbeing (Section 5.3), connection to nature and workplace design (Section 5.4) are also explored. In Section 5.5 a model which considers connection to nature as a resource is proposed.

Singh et al. (2010) [41] investigated the effects of improved indoor environmental quality (IEQ) on perceived health and productivity in occupants who moved from conventional to Leadership in Energy and Environmental Design (LEED)-rated green office buildings. They presented the linkages between seven IEQ attributes and productivity/wellbeing attributes. IEQ attributes were: indoor air quality, temperature, humidity, ventilation, lighting, acoustics, and ergonomic design and safety. Two physical wellbeing attributes used by Singh et al. (2010) [41] were asthma and respiratory allergies. Two psychological wellbeing attributes were stress and depression. They reported perceived improvements in stress and depression after the move into the new LEED rated building. Heerwagen (2000) [42] summarised the potential implications of green building design on health and wellbeing. They reported that increased access to daylight and window views are likely to have positive impacts on psychological functioning and wellbeing. Views with natural settings or urban settings with trees are associated with stress reduction and positive emotional states.

5.1. How Current Green Building Rating Tools Addess Wellbeing and Connection to Nature

The current green building rating tools: Leadership in Energy and Environmental Design (LEED) [43], Building Research Establishment Environmental Assessment Method (BREEAM) [44], Green Star Design and as Built v 1.2 (GBCA, 2017) [45], GreenMark (BCA, 2016) [46] were investigated by Loh (2016) [47] for how the rating tools included the health and wellbeing of the building occupants. Loh (2016) [47] reported that no direct points are awarded to design decisions made specifically for the health and wellbeing of the building occupants (Table 2). It was found that no direct points are awarded for connection to nature by the current green building rating tools.

Table 2. Green building rating tools and ways of addressing wellbeing (adapted from Loh, 2016 [47]).

Tool	Physical and Psychological Wellbeing
LEED—US Leadership in Energy and Environmental Design	Indirectly affected through air, water and light
BREEAM—UK Building Research Establishment Environmental Assessment Method	Items termed for wellbeing are somewhat similar to IEQ items in other rating tools
Green Star—Australia Green Building Council of Australia (GBCA)	Indirectly affected through IEQ
GreenMark—Singapore Building and Construction Authority (BCA)	Indirectly affected through IEQ

5.2. Biophilic Design

The Biophilic design hypothesis posits that there will be an instinctive relationship between humans and a natural ecosystem, a natural attachment to nature [48]. The concept aims to make use of natural elements in architectural and environmental design. Table 3 shows a summary of some of the elements and patterns used in biophilic design.

The traditional workspace should therefore mimic natural environments in three aspects: (a) nature in space which refers to a direct experience of nature such as light or wind or plants (b) the production of nature mimicking nature where it is possible using natural materials or symbolic representation of nature (c) Characteristics of space in which one can observe nature or provision of a space for refuge. Using this principle, it would be easier to design such biophilic features into buildings and work areas before they are built. Incorporating features such as use of natural light or use of natural construction materials may be difficult in renovation projects.

Table 3. Biophilic design patterns (Adapted from Browning, Ryan & Clancy, 2014 [49] and Lee and Park, 2018 [50]).

Category	Design Patterns
Nature in the space patterns	Visual connection with nature Non-visual connection with nature Non-rhythmic sensory stimuli Thermal and airflow variability Presence of water Dynamic and diffuse Light Connection with natural systems
Natural analogues patterns (Production of nature)	Biomorphic forms and patterns Material connection with nature Complexity and order
Nature of the space patterns (Characteristics of space)	Prospect Refuge Mystery Risk/Peril

Kellert and Calabrese (2015) [51] presented three categories of experience of nature in biophilic design framework. They are: direct experience of nature (natural light, air, water, plants, animals, weather, natural landscapes and ecosystems, fire); indirect experience of nature (images of nature, natural materials, natural colours, simulating natural light and air, naturalistic shapes and forms, evoking nature, information richness, age, change, and the patina of time, natural geometries, biomimicry); and experience of spatial features characteristic of the natural environment (prospect and refuge, organized complexity, integration of parts to wholes, transitional spaces, mobility and wayfinding, cultural and ecological attachment to place). The combinations of various biophilic design features on experience of nature would have varying effects on wellbeing. The current literatures have been largely prescriptive about the built environment design patterns enabling connection to nature, they did not however quantify these effects on wellbeing.

5.3. Connection to Nature and Wellbeing

Over the past few decades, researchers in psychology have reported on how nature and green spaces have a positive effect on mental wellbeing, physical wellbeing, vitality, personal autonomy and growth [52]. Ulrich (1979) [53] argued that fundamentally people showed an aesthetic preference for natural landscapes over urban ones and that these natural landscapes have positive influences on emotional and physiological states. Ulrich (1983) [54] also postulated that people's affective response to the natural environment can be learned in that there is post cognitive processing and adaptive behaviour after the initial affective encounter. Although the key visual properties (complexity, texture, depth, etc.) of natural environments influence aesthetic preference, environmental perception is multimodal and not limited to the visual senses. While studies among Western groups have shown preference for natural environments there are similarities with the preferences of different cultures for visual natural environments, Ulrich (1983) [54] recognised that culture plays an important role in determining aesthetic preference. Prominent man-made features however will depress aesthetic preference. These past works positively underline the strong relationship between natural environment and both emotional and physiological wellbeing. It therefore follows that a well- designed work environment with strong connection to nature would have positive effects on people using the workspace.

The restorative effects of green spaces and nature is well known. Ulrich (1983) [54] also suggested that there are psychophysiological restoration effects in visual landscapes and that people recuperates from stress more quickly if they are exposed to visual encounters with nature. In a study of recovery data of patients in a hospital, patients who had views of trees showed a shorter post-operative stay than patients who were assigned rooms with wall views. Views of vegetation and water elicit positive feelings and reduced fear which might foster restoration from stress [55]. Therefore, it is suggested that the immediate work environment with strong connectedness to nature would therefore promote wellbeing by promoting restorative processes.

Workplace design such as offices or workspaces are also very much dependent on the overall building design. Salingaros and Madsen (2008) [56] argued that architects in their design, impose artificial meaning on buildings that do not connect human beings to the place they inhabit and offered a theory to explain how well connected buildings create a sense of wellbeing, with positive and therapeutic consequences on physiology. This connection between nature and wellbeing has been applied to the design of built environment, especially in hospital design. There is a strong body of evidence to suggest that proper hospital designs that provide full views of gardens, access to nature, exposure to art and low noise can be very effective in reducing stress and pain for patients [57].

Cleary et al. [58] defined nature to be an environment free of human interference and suggested there is a continuum of different levels of human intervention in an urban environment such as gardens to urban forests, canals and rivers. The connection to nature is the mix of feelings and attitudes that a person has towards nature and suggests that eudaimonic wellbeing is associated with nature connection. This continuum of different levels of human intervention is an important design consideration for architects, designers and managers.

5.4. Connection to Nature and the Workplace Design

In the design of the urban environment to incorporate elements of nature, even something as simple as indoor plants was found to have a positive effect on workers' wellbeing [59]. On a larger scale, in a longitudinal study of workers in a building site office, Gray and Birrell [60] reported that biophilic designed workspaces have a positive effect on wellbeing, fostering a collaborative work environment and job satisfaction. Yet in another study of 64 knowledge workers, the types of workspaces that are considered conducive to promoting wellbeing were investigated and outdoor workspaces were found to be most conducive (Mangone et al.) [61]. Despite evidence of effectiveness in promoting wellbeing, the use of nature connectedness has not been widely applied to ergonomics or human factors engineering [62] and this calls for a critical review of human factors and ergonomics (HFE) principles to incorporate green principles. Richardson et al. [63] suggested that the three research themes of connectedness to nature, wellbeing and workplace design are closely connected and further research needs to be done to examine and understand how these can be applied. In summary, connection to nature has positive effects on wellbeing and promotes restoration from being stressed. Connection to nature can be incorporated at varying degrees or levels into the local areas such as the immediate workspace as well as integrated into the overall building design.

5.5. Connection to Nature as a Resource

In a worldwide study by the World Health Organization, the Global Burden of Diseases Study [64], cited stress related illness, such as mental health disorders and cardio-vascular disease, to be the two largest contributors to disease. This was enough to call for a worldwide action plan to improve mental health [65]. Schultz (2002) [66] (p. 61) stated "We are borne in nature; our bodies are formed of nature; we live by the rules of nature.". Schultz (2002) [66] argued that people living in industrialised nations are largely segregated from nature. With a diminished connection to nature, the increasing pressure on urban space and the ubiquitous technological presence we have less opportunity to recuperate our mental and physical energy. One of the quotes by Leo Tolstoy, a Russian writer and philosopher, "One of the first conditions of happiness is that the link between Man and Nature shall not be broken." [67] (paragraph. 14) clearly highlights the importance of connectedness to nature. Marcus and Sachs (2013) [68] noted that connection to nature is one of the most effective forms of positive distraction in the healthcare setting. "The deeper the connection to nature is, the grater the therapeutic benefits are." [68].

Based on self-reports of 267 participants in a survey, Tauber (2012) [69] concluded that connection to nature directly affects an individual's physical wellbeing and psychological wellbeing. Connection to nature may also be considered to be a positive stimuli to promote engagement and pleasure for people with dementia [70]. Many researchers [51,52,58–62,71–74] have argued for the wellbeing benefits of connection to nature. The common reasons include health benefits for employees, improved wellbeing, positive feelings of happiness, creativity and productivity. Some have also argued that humans have an innate attraction to nature. Others have argued it provides for sensory richness, natural rhythms, challenges in nature and local distinctiveness. Yet others have argued that channellings the outdoors into our own workspace is within our psyche and that it is natural for humans to work in natural environment research. However, some of these claims are not yet grounded in rigorous research.

Amidst these claims of the health benefits of connection to nature, there is some ongoing academic work to investigate the health effects. The framework of biophilic design presented by Kellert and Calabrese (2015) [51] applies various natural phenomena to the built environment. The Human Spaces report (2016) [75] reported positive interim results of an ongoing study on the use of biophilic designs in workspaces. In a review of the psychological literature on the health and wellbeing benefits of biophilic design, Gillis and Gatersleben (2015) [76] also concluded that the presence of restorative qualities in biophilic design in a built environment can help foster recovery from stress and mental fatigue. The health benefits of helping patients recover is also documented by Totaforti (2018) [77] using the case study of a hospital, where she argued for the use of the healing powers of nature.

Table 4 shows strengths and limitations of the current models and concepts concerning work design, workplace design, stress, wellbeing and connection to nature as put forward over the years. It should be noted that none of these models consider connection to nature in work design. It justified the development of a new improved model which includes connection to nature.

Table 4. Summaries of models and concepts.

Model	Strengths	Limitations
JD-C Karasek [11]	Related work design to occupational stress. Corelated control can alleviate the negative demands of a job.	Simplistic in approach with just two elements. Inconsistent empirical evidence [19,22]. Does not account for any resources.
JD-R Bakker [7]	Correlated job resources to job demand, to explain employee burnout and engagement.	The nature of job demands and resources is not clear or not explained. The role of personal resources is not implicit in the model [26]. Resources are also discussed generally and not in relation to nature.
COR Hobfoll [33]	Described stress as a coping behaviour in which people build or deplete resources.	Does not explain the source of stress or the role of other resources in the workplace.
SDM of stress Morris et al. [36]	Incorporated coping styles and, locus of control and health.	No discussion on workplace design or the role of nature.
SD of workplace stress Jetha et al. [39]	Incorporated workplace stress, job demands, job resources and control as a model using system dynamics approach.	No direct discussion on workplace design or the role of nature.
Nature as a new paradigm Richardson et al. [63]	Integrated the three research themes of connected to nature, workplace design and wellbeing and the role of ergonomics.	The role of stress is not articulated. Not supported by empirical evidence.

Based on the review of the relevant research literature we proposed an improved model. We apply the COR theory [9] and combine it with JD-R model [7] to explain the mechanism by which a job resource of having green natural work environment can reduce stress which in turn in alleviates the negative effect of a high job demand situation (Figure 6). In particular, the combined JD-R model and COR theory is used to show that having connection to nature elements of nature has a positive effect on organizational outcomes. We postulate that personal resources are an important part of the process and that (a) physical resource can directly promote wellbeing. Some may argue that green natural environments are a physical resource; (b) physical resources affect our energies [9] and can be seen as aiding the value of other types of resources. An example of this would be that nature connected spaces maximize our use of relevant personal traits in that it allows concentration and a sense of awareness.

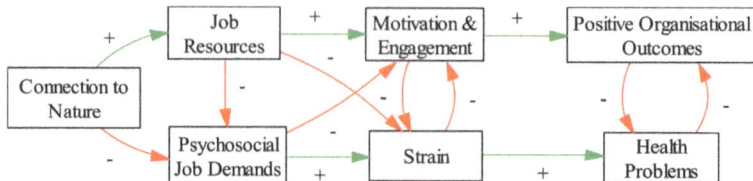

Figure 6. A causal diagram relating connection to natural and wellbeing outcomes. Note: green arrows indicate positive causal relations, and red arrows indicate negative causal relations.

The strength of the model is that it can explain the causal effects of connection to nature as a job resource. The limitation is that it may be challenging to validate the model with objective measures as it requires multi-disciplinary investigations (architectural, building engineering, human factors engineering, psychology, health science and information technology).

The presented model addresses the limitations of the existing models and concepts shown in Table 4. The new conceptual model presented in Figure 6 accounts for job resources, clarifies the role of resources in relation to nature, explains the source of stress (or strain) and its relation to resources, includes a role for connection to nature. It implies that connection to nature must be included as part of good workplace design for wellbeing benefits. The model shows how connection to nature could reduce health problems and promote positive organisational outcomes. The following section is an extensive discussion of the how this model bridges the knowledge gaps.

6. Discussion

6.1. Connection to Nature, Work Design, and Wellbeing

The proposed model incorporates connection to nature as an essential element in workplace design. Previous research has not combined the three elements of connection to nature, workplace design, and wellbeing in a consistent manner [63]. As suggested by Richardson et al. (2017) [63], the consideration of connection to nature introduces a new paradigm in workplace design and wellbeing, two themes that are well developed in the field of ergonomics. Cleary et al. [58] argued that nature exists in the urban environment as urban nature in varying degrees and spans a continuum that correlates to different levels of human influence. This varying of human influence can be categorised by the level of human interference. At low levels, it can be a picture of nature or a potted plant in an office. This can progress to a higher level of connection to nature such as use of furniture made of natural materials or mimicking natural surroundings. At the highest level, this would mean even less artificial nature such as locating the workplace in a garden. The model also suggests that the degree of connection to nature can be measured and used as an element in workplace design. There is an ever-increasing body of research and empirical evidence that connection to nature has positive effects on mental health and wellbeing [52]. The design of a workplace conducive to aiding high performing individuals and teams would be an invaluable resource. While researchers in the field of ergonomics investigated many technical aspects of workplace design such as layout, use of technology, or safety, the consideration of connection to nature would enhance the use of personal resources (psychological and physical wellbeing). Thus, there is a need to integrate the field of ergonomics, building, and workplace design with the concept of connection to nature to support more positive organizational environments.

6.2. Managerial Implications of Connection to Nature Enabling Workspaces

There are two key managerial implications for incorporating connection to nature into workspaces: physical wellbeing and psychological wellbeing of workspace users. In catering for the physical wellbeing, managers and designers ensure that the people using the workspace are safe and not exposed to any form of hazards. The second aspect is important in that psychological wellbeing such as the engagement and positive emotions of workspace users can be enhanced by good workspace design.

6.3. The Cost of Unhealthy Workplaces

Buildings that are poorly designed may lead to occupants displaying symptoms of sick building syndrome, a medical condition where occupants may display a number of symptoms such as general irritation of skin, or general health problems, headaches, and hypersensitivity. A New South Wales Standing Committee on Public Works Report (2001) outlined the widespread problem in Australia and pointed to generally accepted causes of this syndrome that are related to the way buildings are designed, fitted out, or operated. This includes (1) poor building design, particularly when the occupants are totally isolated from the outside environment; (2) use of artificial lighting and 'air'; (3) indoor air pollutants such as chemical, biological, and physical originating from building and fit-out materials and heating ventilation and air conditioning (HVAC) systems; (4) poor design and operation of HVAC systems; and (5) psychosocial factors such as management attitudes in the workplace, stress, and interpersonal relationship. The report recommended the use of sustainable design principles, environmental design considerations, natural ventilation, limiting sources on pollutants, and improved indoor air quality as steps towards minimizing sick building syndrome. These measures are totally in line with the principles of biophilic design of workspaces [49].

6.4. Occupant Safety and Wellbeing

Apart from the sick building syndrome, badly designed workspaces also pose risks to occupants in the form of ergonomic issues and exposure to noise. Ergonomics is the study of human interaction with other elements of a system. Attaianese (2012) [78] applied the same principles to building design and suggested that high performing buildings should take into consideration the elements of (1) energy efficiency with a low carbon footprint, (2) functionality for its planned use by its occupants, (3) ease of operation and maintenance, and (4) protecting occupant comfort. Through its designed characteristics, the building should contribute to sustainable development and at the same time, decrease use of resources; decrease environmental impact; and increase of health, safety, and comfort of the occupants. Protecting occupants from noisy environment and the use of appropriate lighting levels are examples of occupant comfort. These considerations are also in line with the biophilic design principles [49].

6.5. Psychological Wellbeing (Promoting Human Capital Development)

Apart from the physical aspects of wellbeing, connection to nature is also known to give rise to positive emotions. These positive emotions are resources that can be used to create more resources, which in turn is good for the organisation. Human capital is considered one of the most important factors of production. In the 18th century, Adam Smith defined it as the useful benefits from the members of a society. It comprises knowledge, skills, competences, habits, and personalities needed to perform labour. The nature of tasks assigned to working people has changed over the years, from the menial labour during the Industrial Revolution to highly skilled occupations in the 20th century. With the rapid advancement of technology where computers equipped with artificial intelligence can replace human beings in more and more tasks, the work performed by people in the 21st century will change rapidly, and will require higher amounts of creative and innovative outputs from people [79]. Innovation is a nebulous concept, which may mean different things to different people. However, at the root of it, organisations need their people to be creative in doing things differently from others to stay ahead of the competition.

New skills such as the ability to work well in teams will be emphasised. The measure of people's emotional intelligence, the emotional quotient, will be as important as the traditional measure of intelligence, the intelligence quotient (IQ) [80]. In the aim to raise productivity and profitability, organisations will have to be on top of their game and ensure that their working units have high performing teams.

While it is recognised that providing right skills, training, and opportunities is an important step, organisations must also create the comfortable indoor environment and work climate for this to happen. Good workspace design is seen as a facilitator of this aspect of organisational productivity. It enhances people abilities to stay healthy [81] and enhances collaboration and teamwork [82,83]. It also enhances people's's abilities to be innovative through better work engagement [18]. The underlying principle behind Biophilia is that people want and need to be close to nature [48]. Connecting the work to nature would be an enabling factor in work productivity.

7. Conclusions

Provision of connection to nature in workspaces would contribute to a holistic and comprehensive approach in the design and build of workplaces and work environments as it enhances physical and psychological wellbeing, as well as positive emotions at work. A new model that incorporates connection to nature into workplace design for wellbeing has been proposed. There are two types of employee wellbeing: a physical or physiological wellbeing, and mental or psychological wellbeing. The connection to nature in workplaces is a job resource that has this dual effect on wellbeing. The model we have presented combines the three broad research themes of connection to nature, workplace design, and wellbeing, as suggested by Richardson et al. (2017) [63].

The innovative aspects of the model are as follows:

- It includes nature as a resource or a pre-requisite in workplace design.
- It explains the mechanism of how these themes (connection to nature, workplace design, wellbeing) are connected. It is through connection to nature as a job resource that stress is reduced (wellbeing improves). It should be noted that previous research has either focused on connection to nature and workplace design or connection to nature and wellbeing, or workplace design and wellbeing, but not all three combined. Only Richardson et al. (2017) [63] showed that the three broad research themes of workplace design, wellbeing, and connection to nature can be combined in a new ergonomic paradigm.

We propose connection to nature in workplace design as a spectrum. We attempt to organise various connection to nature themes into a spectrum of categories. This spectrum illustrates various levels of connection to nature ranging from low levels, for instance, having a photograph of a plant, to mid levels, for example, the use of natural materials like wood and stone, and up to high levels such as locating an office in a forest or a zoo.

8. Future Research

1. We recognise that temporal and spatial aspects (e.g., frequency and duration) of connection to nature has varying effects on people. This needs to be explored further.
2. There is a need to refine the subjective measures of wellbeing and measures of workplace design that relate with connection to nature. Earlier research applied subjective self-reported responses to surveys, however, with the current advances of the sensor technology and the Internet of Things (IoT), wellbeing can be objectively measured.
3. Guidelines on the inclusion of connection to nature in workplace and building design considering human factors need to be developed.

Author Contributions: F.K.P.H. drafted the manuscript and critically reviewed the intellectual contents; L.A. conceived this investigation, developed all figures using Vensim PLE [84] and substantively revised the draft. Both have approved the final published version.

Funding: This research received no external funding.

Acknowledgments: The authors acknowledged Dr Sally Wilson, Department of Infrastructure Engineering, the University of Melbourne for proof reading the manuscript and improving the language aspects. Many thanks are due to the editor: Dr Zhonghua Gou, Griffith School of Environment for inviting us to contribute.

Conflicts of Interest: The authors declare no conflict of interest.

References

1. WGBC—World Green Building Council. *Health, Wellbeing and Productivity in Offices: The Next Chapter for Green Building*; World Green Building Council: London, UK, 2014.

2. Stokols, D. Chapter 34: Psychosocial and Organizational Factors. In *The ILO Encyclopaedia of Occupational Health & Safety*, 4th ed.; Stellman, J.M., Ed.; International Labor Organization: Geneva, Switzerland, 2011; Volume 4.

3. Higginbottom, K. Workplace Stress Leads to Less Productive Employees. Available online: https://www.forbes.com/sites/karenhigginbottom/2014/09/11/workplace-stress-leads-to-less-productive-employees/#3b9b276931d1 (accessed on 24 July 2018).

4. Houtman, I.; Jettinghof, K.; Cedillo, L. *Raising Awareness of Stress at Work in Developing Countries*; World Health Organization: Geneva, Switzerland, 2007.

5. Keser, S. Occupational Stress and Depression: Insights from Traditional and Emerging Views. Ph.D. Thesis, Australian National University, Canberra, Australia, 2016.

6. Sauter, S.; Murphy, L.; Colligan, M.; Swanson, N.; Hurrell, J.; Scharf, F.; Sinclair, R.; Grubb, P.; Goldenhar, L.; Alterman, T. *Stress at Work*; DHHS (NIOSH) Publication; National Institute for Occupational Safety and Health: Cincinnati, OH, USA, 1999; pp. 1–25.

7. Bakker, A.B.; Demerouti, E. The job demands-resources model: State of the art. *J. Manag. Psychol.* **2007**, *22*, 309–328. [CrossRef]

8. Hobfoll, S.E. Social and psychological resources and adaptation. *Rev. Gen. Psychol.* **2002**, *6*, 307–324. [CrossRef]

9. Hobfoll, S.E. Conservation of resources: A new attempt at conceptualizing stress. *Am. Psychol.* **1989**, *44*, 513–524. [CrossRef] [PubMed]

10. Clarivate. Web of Science. Available online: www.webofknowledge.com/ (accessed on 1 July 2018).

11. Karasek, R.A., Jr. Job demands, job decision latitude, and mental strain: Implications for job redesign. *Admin. Sci. Q.* **1979**, 285–308. [CrossRef]

12. Demerouti, E.; Bakker, A.B.; Nachreiner, F.; Schaufeli, W.B. The job demands-resources model of burnout. *J. Appl. Psychol.* **2001**, *86*, 499–512. [CrossRef] [PubMed]

13. Xanthopoulou, D.; Bakker, A.B.; Demerouti, E.; Schaufeli, W.B. The role of personal resources in the job demands-resources model. *Int. J. Stress Manag.* **2007**, *14*, 121–141. [CrossRef]

14. Mayerl, H.; Stolz, E.; Waxenegger, A.; Rásky, É.; Freidl, W. The role of personal and job resources in the relationship between psychosocial job demands, mental strain, and health problems. *Front. Psychol.* **2016**, *7*, 1214. [CrossRef] [PubMed]

15. Hussung, T. Understanding Work Stress: Causes, Symptoms and Solutions. Available online: https://online.csp.edu/blog/business/understanding-work-stress (accessed on 5 August 2018).

16. LaMontagne, A.; Keegel, T. *Reducing Stress in the Workplace: An Evidence Review: Full Report*; Victorian Health Promotion Foundation: Melbourne, Australia, 2012.

17. Harvey, S.B.; Joyce, S.; Tan, L.; Johnson, A.; Nguyen, H.; Modini, M.; Groth, M. *Developing a Mentally Healthy Workplace: A Review of the Literature*; National Mental Health Commission and Mentally Healthy Workplace Alliance: Sydney, Australia, 2014.

18. Hui, K.P. *A Model of Goal Orientation, Work Engagement, the Need for Achievement and Innovation*; University of Newcastle: Callaghan, Australia, 2013.

19. De Lange, A.H.; Taris, T.W.; Kompier, M.A.; Houtman, I.L.; Bongers, P.M. "The very best of the millennium": Longitudinal research and the demand-control-(support) model. *J. Occup. Health Psychol.* **2003**, *8*, 282–305. [CrossRef] [PubMed]

20. Brandtzæg, P.B.; Følstad, A.; Heim, J. Enjoyment: Lessons from Karasek. In *Funology 2*; Springer: Cham, Switzerland, 2018; pp. 331–341.

21. Wallgren, L.G.; Hanse, J.J. Job characteristics, motivators and stress among information technology consultants: A structural equation modeling approach. *Int. J. Ind. Ergon.* **2007**, *37*, 51–59. [CrossRef]

22. Van der Doef, M.; Maes, S. The job demand-control (-support) model and psychological well-being: A review of 20 years of empirical research. *Work Stress* **1999**, *13*, 87–114. [CrossRef]

23. Johnson, J.V.; Hall, E.M. Job strain, work place social support, and cardiovascular disease: A cross-sectional study of a random sample of the Swedish working population. *Am. J. Public Health* **1988**, *78*, 1336–1342. [CrossRef] [PubMed]

24. Demerouti, E.; Bakker, A.B.; De Jonge, J.; Janssen, P.P.; Schaufeli, W.B. Burnout and engagement at work as a function of demands and control. *Scand. J. Work Environ. Health* **2001**, *27*, 279–286. [CrossRef] [PubMed]

25. Landsbergis, P.A.; Schnall, P.L.; Deitz, D.; Friedman, R.; Pickering, T. The patterning of psychological attributes and distress by "job strain" and social support in a sample of working men. *J. Behav. Med.* **1992**, *15*, 379–405. [CrossRef] [PubMed]

26. Schaufeli, W.B.; Taris, T.W. Chapter 4: A critical review of the job demands-resources model: Implications for improving work and health. In *Bridging Occupational, Organizational and Public Health*; Springer: Dordrecht, The Netherlands, 2014; pp. 43–68.

27. Salanova, M.; Agut, S.; Peiró, J.M. Linking organizational resources and work engagement to employee performance and customer loyalty: The mediation of service climate. *J. Appl. Psychol.* **2005**, *90*, 1217–1227. [CrossRef] [PubMed]

28. Hakanen, J.J.; Koivumäki, J. Engaged or exhausted—How does it affect dentists' clinical productivity? *Burnout Res.* **2014**, *1*, 12–18. [CrossRef]

29. Llorens-Gumbau, S.; Salanova-Soria, M. Loss and gain cycles? A longitudinal study about burnout, engagement and self-efficacy. *Burnout Res.* **2014**, *1*, 3–11. [CrossRef]

30. Zeidner, M.; Ben-Zur, H.; Reshef-Weil, S. Vicarious life threat: An experimental test of conservation of resources (COR) theory. *Personal. Individ. Differ.* **2011**, *50*, 641–645. [CrossRef]

31. Antonovsky, A. *Health, Stress, and Coping*; Jossey-Bass Inc.: San Francisco, CA, USA, 1979.

32. Wellman, B. Applying network analysis to the study of support. In *Social Networks and Social Support*; Gottlieb, B.H., Ed.; Sage Publications: Beverly Hills, CA, USA, 1981; pp. 171–200.

33. Hobfoll, S.E.; Johnson, R.J.; Ennis, N.; Jackson, A.P. Resource loss, resource gain, and emotional outcomes among inner city women. *J. Personal. Soc. Psychol.* **2003**, *84*, 632–643. [CrossRef]

34. Xanthopoulou, D.; Bakker, A.B.; Demerouti, E.; Schaufeli, W.B. Reciprocal relationships between job resources, personal resources, and work engagement. *J. Vocat. Behav.* **2009**, *74*, 235–244. [CrossRef]

35. Harter, J.K.; Schmidt, F.L.; Hayes, T.L. Business-unit-level relationship between employee satisfaction, employee engagement, and business outcomes: A meta-analysis. *J. Appl. Psychol.* **2002**, *87*, 268–279. [CrossRef] [PubMed]

36. Morris, A.; Ross, W.; Ulieru, M. A system dynamics view of stress: Towards human-factor modeling with computer agents. In Proceedings of the 2010 IEEE International Conference on Systems, Man and Cybernetics, Istanbul, Turkey, 10–13 October 2010; pp. 4369–4374.

37. Hobfoll, S.E. *The Ecology of Stress*; Hemisphere Pub. Corp.: New York, NY, USA, 1988.

38. Simgua. A Next-Generation, No-Nonsense Modeling Application That Helps You Develop Powerful Models and Simulations. Available online: http://simgua.com/ (accessed on 25 July 2018).

39. Jetha, A.; Kernan, L.; Kurowski, A. Conceptualizing the dynamics of workplace stress: A systems-based study of nursing aides. *BMC Health Services Res.* **2017**, *17*, 12. [CrossRef] [PubMed]

40. Ventana Systems. Vensim. Available online: http://vensim.com/ (accessed on 25 July 2018).

41. Singh, A.; Syal, M.; Grady, S.C.; Korkmaz, S. Effects of green buildings on employee health and productivity. *Am. J. Public Health* **2010**, *100*, 1665–1668. [CrossRef] [PubMed]

42. Heerwagen, J. Green buildings, organizational success and occupant productivity. *Build. Res. Inf.* **2000**, *28*, 353–367. [CrossRef]

43. USGBC—U.S. Green Building Council. Leadership in Energy and Environmental Design (LEED) v.4. Available online: https://new.usgbc.org/leed-v4 (accessed on 4 August 2018).

44. BRE—Building Research Establishment. Building Research Establishment Environmental Assessment Method (BREEAM). Available online: http://www.breeam.com/ (accessed on 4 August 2018).

45. GBCA—Green Building Council of Australia. Green Star Design and as Built v 1.2. Available online: https://new.gbca.org.au/green-star/rating-system/design-and-built/ (accessed on 4 August 2018).

46. BCA—Building & Construction Authority. BCA Green Mark Assessment Criteria and Application Forms. Available online: https://www.bca.gov.sg/greenmark/green_mark_criteria.html (accessed on 4 August 2018).

47. Loh, S. Inclusion of well-being and ecological values in green building assessment tools. In Proceedings of the International Conference on Architecture, Landscape, Interior Design & Urban Planning (ALIU 2016), Kuala Lumpur, Malaysia, 8 December 2016; pp. 58–62.

48. Ulrich, R.S.; Simons, R.F.; Losito, B.D.; Fiorito, E.; Miles, M.A.; Zelson, M. Stress recovery during exposure to natural and urban environments. *J. Environ. Psychol.* **1991**, *11*, 201–230. [CrossRef]

49. Browning, W.D.; Ryan, C.O.; Clancy, J.O. *14 Patterns of Biophilic Design: Improving Health and Well-Being in the Built Environment*; Terrapin Bright Green, LLC: New York, NY, USA, 2014.

50. Lee, H.C.; Park, S.J. Assessment of importance and characteristics of biophilic design patterns in a children's library. *Sustainability* **2018**, *10*, 987. [CrossRef]

51. Kellert, S.; Calabrese, E. The Practice of Biophilic Design. 2015, pp. 1–25. Available online: www.biophilic-design.com (accessed on 23 September 2018).

52. Passmore, H.-A. Feeling blue-get green. *Earth Common J.* **2011**, *1*, 80–86.

53. Ulrich, R.S. Visual landscapes and psychological well-being. *Lands. Res.* **1979**, *4*, 17–23. [CrossRef]

54. Ulrich, R.S. Aesthetic and affective response to natural environment. In *Behavior and the Natural Environment*; Altman, I., Wohlwill, J.F., Eds.; Springer: Boston, MA, USA, 1983; pp. 85–125.

55. Ulrich, R.S. View through a window may influence recovery from surgery. *Science* **1984**, *224*, 420–421. [CrossRef] [PubMed]

56. Salingaros, N.; Masden, K. Neuroscience, the natural environment, and building design. In *Biophilic Design: The Theory, Science and Practice of Bringing Buildings to Life*; John Wiley: New York, NY, USA, 2008; pp. 59–83.

57. Ulrich, R.S.; Zimring, C.; Quan, X.; Joseph, A. The environment's impact on stress. In *Improving Healthcare with Better Building Design*; ACHE Management Series; Health Administration Press: Baltimore, MD, USA, 2006; pp. 37–61.

58. Cleary, A.; Fielding, K.S.; Bell, S.L.; Murray, Z.; Roiko, A. Exploring potential mechanisms involved in the relationship between eudaimonic wellbeing and nature connection. *Landsc. Urban Plan.* **2017**, *158*, 119–128. [CrossRef]

59. Kalantzis, A. The Impact of Indoor Plants on Well-Being in the Workplace. Master Thesis, University of the Witwatersrand, Johannesburg, South Africa, 2016.

60. Gray, T.; Birrell, C. Are biophilic-designed site office buildings linked to health benefits and high performing occupants? *Int. J. Environ. Res. Public Health* **2014**, *11*, 12204–12222. [CrossRef] [PubMed]

61. Mangone, G.; Capaldi, C.A.; van Allen, Z.M.; Luscuere, P.G. Bringing nature to work: Preferences and perceptions of constructed indoor and natural outdoor workspaces. *Urban For. Urban Green.* **2017**, *23*, 1–12. [CrossRef]

62. Lumber, R.; Richardson, M.; Albertsen, J.-A. Hfe in biophilic design: Human connections with nature. In *Ergonomics and Human Factors for a Sustainable Future: Current Research and Future Possibilities*; Thatcher, A., Yeow, P.H.P., Eds.; Springer: Singapore, 2018; pp. 161–190.

63. Richardson, M.; Maspero, M.; Golightly, D.; Sheffield, D.; Staples, V.; Lumber, R. Nature: A new paradigm for well-being and ergonomics. *Ergonomics* **2017**, *60*, 292–305. [CrossRef] [PubMed]

64. WHO—World Health Organization. *The World Health Report 2002: Reducing Risks, Promoting Healthy Life*; World Health Organization: Geneva, Switzerland, 2002.

65. WHO—World Health Organization. *Mental Health Action Plan 2013–2020*; World Health Organization: Geneva, Switzerland, 2013.

66. Schultz, P.W. Chapter 4: Inclusion with nature: The psychology of human-nature relations. In *Psychology of Sustainable Development*; Schmuck, P., Schultz, W.P., Eds.; Springer: Boston, MA, USA, 2002; pp. 61–78.

67. Bright Side. 20 Quotes from Leo Tolstoy That Reveal the Secret to a Happy Life. Available online: https://brightside.me/wonder-people/20-quotes-from-leo-tolstoy-that-reveal-the-secret-to-a-happy-life-145005/ (accessed on 4 August 2018).

68. Marcus, C.C.; Sachs, N.A. *Therapeutic Landscapes: An Evidence-Based Approach to Designing Healing Gardens and Restorative Outdoor Spaces*; John Wiley & Sons: Hoboken, NJ, USA, 2013.

69. Tauber, P.G. An Exploration of the Relationships among Connectedness to Nature, Quality of Life, and Mental Health. Master's Thesis, Utah State University, Logan, UT, USA, 2012.

70. Chau, H.-W.; Newton, C.; Woo, C.; Ma, N.; Wang, J.; Aye, L. Design lessons from three Australian dementia support facilities. *Buildings* **2018**, *8*, 67. [CrossRef]

71. Hedges, J. How to get kids offline, outdoors, and connecting with nature: 200+ creative activities to encourage self-esteem, mindfulness, and wellbeing. *Health Soc. Work* **2015**, *40*, 158–160. [CrossRef]

72. Nousiainen, M.; Lindroos, H.; Heino, P. *Restorative Environment Design*; Series A. Nr 76; Kymenlaakso University of Applied Sciences Publications: Kuvola, Finland, 2016.

73. Haluza, D.; Simic, S.; Höltge, J.; Cervinka, R.; Moshammer, H. Connectedness to nature and public (skin) health perspectives: Results of a representative, population-based survey among Austrian residents. *Int. J. Environ. Res. Public Health* **2014**, *11*, 1176–1191. [CrossRef] [PubMed]

74. Russell, R.; Guerry, A.D.; Balvanera, P.; Gould, R.K.; Basurto, X.; Chan, K.M.A.; Klain, S.; Levine, J.; Tam, J. Humans and nature: How knowing and experiencing nature affect well-being. *Annu. Rev. Environ. Resour.* **2013**, *38*, 473–502. [CrossRef]

75. Cooper, C.; Browning, B. The Global Impact of Biophilic Design in the Workplace. 2015. Available online: http://architecturenow.co.nz/articles/the-global-impact-of-biophilic-design-in-the-workplace/ (accessed on 4 August 2018).

76. Gillis, K.; Gatersleben, B. A review of psychological literature on the health and wellbeing benefits of biophilic design. *Buildings* **2015**, *5*, 948–963. [CrossRef]

77. Totaforti, S. Applying the benefits of biophilic theory to hospital design. *City Territ. Archit.* **2018**, *5*, 1–9. [CrossRef]

78. Attaianese, E. A broader consideration of human factor to enhance sustainable building design. *Work* **2012**, *41*, 2155–2159. [PubMed]

79. King, E.; Rogers, F. *Intelligence, Personality, and Creativity: Unleashing the Power of Intelligence and Personality Traits to Build a Creative and Innovative Economy*; The World Bank: Washington, DC, USA, 2014.

80. Kim, J.Y. *Human Capital in the 21st Century*; World Bank: Washington, DC, USA, 2014.

81. Miller, N.; Pogue, D.; Gough, Q.; Davis, S. Green buildings and productivity. *J. Sustain. Real Estate* **2009**, *1*, 65–89.

82. Gensler. *What We've Learned about Focus in the Workplace*; Gensler: San Francisco, CA, USA, 2012; pp. 1–8.

83. Gensler. *2013 U.S. Workplace Survey: Key Findings*; Gensler: San Francisco, CA, USA, 2013; pp. 1–28.

84. Ventana Systems. Vensim® Personal Learning Edition. Available online: https://vensim.com/vensim-personal-learning-edition/ (accessed on 4 August 2018).

 © 2018 by the authors. Licensee MDPI, Basel, Switzerland. This article is an open access article distributed under the terms and conditions of the Creative Commons Attribution (CC BY) license (http://creativecommons.org/licenses/by/4.0/).

Article

The Role of Personal Control in Alleviating Negative Perceptions in the Open-Plan Workplace

Zhonghua Gou *, Jian Zhang and Leigh Shutter

School of Engineering and Built Environment, Griffith University, Gold Coast, QLD 4215, Australia;
jian.zhang@griffithuni.edu.au (J.Z.); l.shutter@griffith.edu.au (L.S.)
* Correspondence: z.gou@griffith.edu.au or gouzhonghua@gmail.com; Tel.: +61-7-5552-9510

Received: 26 July 2018; Accepted: 13 August 2018; Published: 14 August 2018

Abstract: Today's office buildings adopt open-plan settings for collaboration and space efficiency. However, the open plan setting has been intensively criticized for its adverse user experiences, such as noise, privacy loss, and over cooling. The provision of personal control in open-plan work environments is an important means to alleviating the adverse perceptions. This research is to investigate the relationship between the availability of personal controls and the degree of control over the physical environment, as well as their effectiveness in alleviating adverse perceptions in open-plan workplaces. The study combined three systematic occupant survey tools and collected responses from open-plan offices in Shenzhen, China. Specifically, this survey covered 12 personal controls in open-plan workplaces; respondents were asked to report their degree of control over the physical environment and also were required to report if they had adverse perceptions such as sick building syndrome in their offices. The results showed that most of the 12 personal controls supported perceived degree of control over the physical environment but only half of them were negatively associated with adverse perceptions. Non-mechanical controls, such as windows and blinds, were found to be more effective than mechanical controls such as fans and air-conditioning in alleviating adverse perceptions. Conflicts were found between task/desk lights and other personal controls. The research generates important evidence for the interior design of open-plan offices.

Keywords: open-plan workplace; environmental control; productivity; satisfaction

---

1. Introduction

Open plan has become the most popular workplace setting for contemporary office buildings. It has many advantages in terms of flexibility, space efficiency, interaction, and collaboration. On the other side, it has been intensively criticized for a multitude of adverse perceptions being experienced by its occupants, such as loss of privacy, loss of identity, low work productivity, various health issues, overstimulation, and low job satisfaction [1]. Human oriented design should be addressed in the design of open-plan workplaces, as employees' health and productivity are associated with a significant portion of business costs. How to alleviate adverse perceptions and promote a positive user experience in open plan offices has become the key issue for workplace design [2].

Previous research on open-plan offices highlighted the role of individual control in alleviating the negative perceptions and promoting human-oriented workplace design [3,4]. The research found that one of key problems of open plan settings is that the freedom of choices is lacking and, therefore, building occupants feel powerlessness and unhappiness, which would consequently decrease task performance [5–10]. Occupants who perceived their control opportunities as being insufficient were less tolerant of their thermal conditions [11] and less stimulated [12]. The provision of personal control in open-plan work environments is one means of preventing the detrimental effects and leading to desirable outcomes [13,14]. A positive association was found between high work control and job satisfaction, work performance and psychological well-being [15,16].

In addition to comfort and satisfaction, energy saving reasons also supported the provision of individual control in office environments. Research demonstrated great energy saving potential due to lighting and ventilation controls [17]. Both laboratory and field experiments suggested that occupants' behaviors, especially how they interacted with personal control, should be included in the loop of control strategies for office buildings [18,19].

Meanwhile, the precedent research also raised some unsolved questions about personal controls in workplaces. For example, how occupants perceive different controls (on lighting, ventilation, noise, etc.)? Whether occupants really use these controls? How these controls work synergistically in workplace environments? How to design these controls to optimize their benefits? This article continues the dialogue on personal controls in open plan office environments to explore the role of different personal controls in open plan settings, aiming to provide evidence and guidelines for human-oriented workplace design.

2. Brief Literature Review

The personal control at workplaces is usually investigated differently in the relevant literature. In some studies [15,20], personal control refers to the degree to which employees perceive they can change their physical work environment, especially by determining, altering, or modifying work areas as necessary to support or to allow their work behaviors. Therefore, the questions were asked in the following ways: "What is your ability to alter physical conditions in your work area?"; "To what degree you feel control over the thermal environment in your workspace?" and the like. On the other hand, in some studies [21,22], personal control is defined in terms of specific environmental adjustment referring to modifying the surroundings themselves, such as opening/closing windows or shades, turning on fans or heating, air diffusers, light switches, and so on. Based on a database accumulated from several recent surveys of office buildings located in a temperate climate, Andersen [23] found that degree of control satisfaction with the perceived control was more likely to affect the prevalence of adverse perceptions and symptoms than the actual control; the most important control actions were access to a thermostat or an operable window. To study personal control, Paciuk [24] identified three dimensions: available control, exercised control, and perceived control and found that perceived degree of control was one of the strongest predictors of thermal comfort and had a significant impact on both comfort and satisfaction.

Noticeably, there is a research gap between actual control opportunities and perceived degree of control over the physical environment. This research is to investigate the availability of control opportunities (windows, blinds, switches, and so on) and the degree of control effectiveness over the physical environment (thermal, lighting, and noise). The research is also to find their relations to the adverse environmental perceptions. Therefore, there are two key research questions: (1) what is the relationship between the actual control opportunities and the degree of control over the physical environment? Which individual controls are most effective in reducing occupants' adverse perceptions in open-plan office environments?

3. Method

Personal control opportunities in workplaces are dependent on a variety of building features, including the windows, blinds, task lights, electrical fans, and the like. In the European Union-funded project Smart Controls and Thermal Comfort (SCATs), McCartney and Nicol identified possible control opportunities [25]: "Open or close a window", "Adjust curtains or a blind", "Open or close an internal door", "Open or close an external door", "Adjust a thermostat", "Adjust a local heater/radiator", "Turn lighting on or off (your desk only), "Turn office lighting on or off", "Adjust the office lighting level (dimmer switch)", "Adjust office air-conditioning", and "Adjust a local fan/air outlet". In the study of Occupant Indoor Environmental Quality (IEQ) Survey by the Center for the Built Environment of UC Berkeley, individual control opportunities in modern office environments, especially on thermal and lighting environments, were identified as follows [26]: "Window blinds or shades", "Operable window", "Thermostat", "Portable heater", "Permanent heater", "Room air-conditioning unit", "Portable fan", "Ceiling fan", "Adjustable air vent in wall or ceiling", "Adjustable floor air vent (diffuser)", "Door to interior space", "Door to exterior space", "Light switch", "Light dimmer", "Window blinds or shades", "Desk (task) light". Combining these two studies, this research listed 12 possible control opportunities in the survey: (1) "Window blinds or shades", (2) "Operable window", (3) "Thermostat", (4) "Room air-conditioning unit", (5) "Ceiling fan", (6) "Portable fan", (7) "Adjustable air vent", (8) "Heater", (9) "Door to interior or exterior space", (10) "Light switch", (11) "Light dimmer", and (12) "Desk (task) light".

For the degree of control over the physical environment and adverse perceptions, the survey used the BUS (Building Use Studies) questionnaire. Respondents were asked to answer the question: "How much control do you personally have over the following aspects of your indoor working environment?" The aspects covered heating, cooling, ventilation, lighting, and noise. The answer ranged from 1 "Little Control" to 7 "Full Control" on each aspect. Respondents were also asked to report their adverse perceptions experience on five aspects representing sick building syndrome: "Do you have any symptoms (see below) which you feel may be associated with being in the building? (We are thinking of any of the following which may appear when you come into the building and disappear when you leave.)" The five aspects covered eyes (irritated, itching, dry, watering), nose (irritated, itching, runny, dry, blocked), throat (sore, constricted, dry mouth), head (headache, lethargy, irritability, difficulty in concentrating), and skin (dryness, itching, irritation, rashes).

To answer the research questions and find out effective personal control resolutions for designing building and interior elements, the research surveyed 411 occupants working in open-plan office settings in six office buildings in Shenzhen (Table 1). To conduct the statistical analyses, a certain ratio of questions to responses (1:10) are needed. In this case, 24 questions (14 on available control opportunities, five on perceived degree of control, and five on sick building syndrome) were asked and the analysis needs at least 240 responses. Therefore, the 411 responses are sufficient for the data analysis. Among the surveyed six office buildings, two buildings use split air-conditioning or room air-conditioning systems which are usually accompanied with ceiling fans and operable windows for alternative ventilation; three use central air-conditioning systems without operable windows; one uses mixed-mode ventilation with both central air-conditioning as well as operable windows. Some of workstations in these buildings also have other control opportunities such as desk lights and portable fans for individual uses. Most of respondents had worked in these building for more than one year. They worked 5.1 days per week on average and 8.3 h per day in their offices. Forty-one percentage respondents perceived that they were seated next to windows while 59% were sitting far from windows.

Table 1. Surveyed office buildings and occupants.

No.	Building's Basic Features	Samples	Demographics
1	4 storeys; split air-conditioning; fans	55	
2	5 storeys; split air-conditioning; fans	33	
3	5 storeys; central air-conditioning	46	60% male & 40% female; 58% under 30 years old & 42% 30 and above year old
4	10 storeys; central air-conditioning	160	
5	4 storeys; central air-conditioning	61	
6	3 storeys; mixed-mode ventilation	56	

4. Results

Frequencies of personal controls available are shown in Table 2. The most frequently reported control opportunity was light switches, and the second was operable windows and window blinds or shades; the least frequently reported was ceiling fans. Numbers of control opportunities available to the respondents are shown in Table 3. The numbers of respondents decreased stably as the numbers of control opportunities increased. Less than half of the respondents had three or more control opportunities available in their workspaces. Table 4 shows the differences in total numbers on each control opportunity. For example, respondents who had window blinds or shades had, on average, a greater number of control opportunities. The most significant difference was found on light dimmers while the least was found on desk/task lights. In other words, light dimmers were more likely to appear with other control opportunities while desk/task lights were less likely to do so. Table 5 shows responses on the degree of control over heating, cooling, ventilation, lighting, and noise. On average, respondents had the highest degree of control over lighting, and then cooling; they had the least degree of control over noise and heating.

Table 2. Frequencies of control opportunities.

C1	Window shades	234
C2	Operable window	241
C3	Thermostat	102
C4	Room air-conditioning unit	119
C5	Ceiling fan	21
C6	Portable fan	75
C7	Adjustable air vent	109
C8	Heater	25
C9	Door to interior or exterior space	137
C10	Light switch	324
C11	Light dimmer	76
C12	Desk (task) light	147
C13	None of the above	98
C14	Others	18

Table 3. Numbers of control opportunities.

Number	Percent	Cumulative Percent
0	21.5	21.5
1	18.6	40.0
2	18.1	58.1
3	17.9	76.1
4	11.7	87.7
5	5.7	93.4
6	3.8	97.2
7	0.6	97.9
8	0.2	98.0
9	0.6	98.6
10	0.6	99.2
11	0.5	99.7
12	0.3	100.0

Table 4. Cross-table of different control opportunities and total numbers of controls.

Control Opportunities (0 = No; 1 = Yes)			Numbers of Controls (0–12)		
			Mean	Std. Deviation	Mean Difference
C1	Window shades	0	1.42	1.36	−2.51
		1	3.94	2.15	
C2	Operable window	0	1.42	1.31	−2.51
		1	3.94	2.10	
C3	Thermostat	0	2.00	1.68	−2.08
		1	4.08	2.94	
C4	Room air-conditioning unit	0	1.89	1.59	−2.38
		1	4.27	2.78	
C5	Ceiling fan	0	2.23	1.86	−2.86
		1	5.10	4.73	
C6	Portable fan	0	2.06	1.76	−2.31
		1	4.37	2.98	
C7	Adjustable air vent	0	1.96	1.64	−2.19
		1	4.15	2.89	
C8	Heater	0	2.27	1.96	−1.49
		1	3.76	3.79	
C9	Door to interior or exterior space	0	1.75	1.48	−2.74
		1	4.49	2.52	
C10	Light switch	0	1.24	1.37	−2.18
		1	3.42	2.08	
C11	Light dimmer	0	2.03	1.67	−2.57
		1	4.59	3.15	
C12	Desk (task) light	0	2.08	1.85	−1.10

Table 5. Degree of control (1 = Little control; 7 = Full control).

Personal Control	Minimum	Maximum	Mean	Median	Std. Deviation
Control Over Heating	1	7	3.86	4	1.85
Control Over Cooling	1	7	4.67	5	1.81
Control Over Ventilation	1	7	4.25	4	1.963
Control Over Lighting	1	7	5.02	5	1.662
Control Over Noise	1	7	3.77	4	1.953

t-test was conducted to see whether perceived degree of control was significantly different between those who had more opportunities and who had fewer opportunities. Table 6 is the *t*-test table for control opportunities and the degree of control. Most of the control opportunities contributed to significant differences on control degree perceptions. However, respondents who had desk or task lights available did not perceive that they had significantly more control over the physical environment including heating, cooling, ventilation, and lighting. Table 7 is the *t*-test table for control opportunities and adverse perceptions. Respondents who had window blinds or shades, operable windows, thermostats, a light switch, a light dimmer or desk (task) lights reported significantly fewer adverse perceptions than those who did not have them. On the other side, room air-conditioning units, ceiling fans, portable fans, adjustable air vents, heater, and doors to interior or exterior space might not significantly contribute to the alleviation of adverse perceptions.

Table 6. Control opportunities and degree of control.

Control Opportunities (0 = No; 1 = Yes)		Control Over Heating (1–7)		Control Over Cooling (1–7)		Control Over Ventilation (1–7)		Control Over Lighting (1–7)		Control Over Noise (1–7)	
		Mean	Diff. (Sig.)	Mean	Diff. (Sig.)	Mean	Diff. (Sig.)	Mean	Diff. (Sig.)	Mean	Diff. (Sig.)
Window shades	0	3.8		4.3		3.7		4.7		3.9	
	1	4.1	−0.4 (0.045)	5.3	−1.0 0.000	5.2	−1.4 0.000	5.6	−0.9 0.000	3.5	0.5 0.004
Operable window	0	3.8		4.3		3.8		4.7		3.9	
	1	4.1	−0.3 0.085	5.2	−0.9 0.000	5.1	−1.3 0.000	5.6	−0.9 0.000	3.5	0.5 0.004
Thermostat	0	3.7		4.5		4.0		4.9		3.7	
	1	5.0	−1.3 0.000	5.6	−1.1 0.000	5.3	−1.3 0.000	5.7	−0.9 0.000	4.3	−0.7 0.002
Room air-conditioning unit	0	3.7		4.5		4.2		4.9		3.6	
	1	4.5	−0.8 0.001	5.3	−0.8 0.000	4.7	−0.5 0.013	5.5	−0.6 0.000	4.6	−1.0 0.000
Ceiling fan	0	3.8		4.6		4.2		5.0		3.7	
	1	6.0	−2.3 0.000	6.0	−1.4 0.001	6.3	−2.2 0.000	6.3	−1.3 0.000	5.9	−2.2 0.000
Portable fan	0	3.7		4.6		4.2		5.0		3.6	
	1	4.7	−1.0 0.001	5.2	−0.6 0.014	4.5	−0.3 0.228	5.5	−0.5 0.013	5.0	−1.3 0.000
Adjustable air vent	0	3.7		4.5		4.1		4.9		3.6	
	1	4.8	−1.1 0.000	5.5	−0.9 0.000	5.1	−1.0 0.000	5.6	−0.7 0.000	4.6	−1.0 0.000
Heater	0	3.8		4.6		4.2		5.0		3.7	
	1	6.0	−2.3 0.000	6.3	−1.7 0.000	6.0	−1.9 0.000	6.3	−1.4 0.000	5.9	−2.2 0.000
Door to interior or exterior space	0	3.7		4.5		4.1		4.8		3.6	
	1	4.5	−0.8 0.000	5.2	−0.7 0.000	4.8	−0.7 0.000	5.7	−0.9 0.000	4.4	−0.9 0.000
Light switch	0	3.7		4.3		3.8		4.5		3.9	
	1	4.0	−0.3 0.042	5.0	−0.7 0.000	4.7	−0.8 0.000	5.5	−1.0 0.000	3.7	0.2 0.139
Light dimmer	0	3.7		4.5		4.1		4.9		3.6	
	1	4.9	−1.2 0.000	5.6	−1.0 0.000	5.3	−1.2 0.000	5.7	−0.7 0.000	4.9	−1.3 0.000
Desk (task) light	0	3.9	0.1 0.675	4.6	−0.2 0.279	4.2	−0.3 0.170	5.0	−0.3 0.096	3.6	−0.5 0.003

Table 8 shows relationships between the total number of control opportunities, degree of control over the physical environment, and adverse perceptions. The perceived degree of control over each indoor physical environment aspect such as heating, cooling, ventilation, lighting, and noise were closely related to each other. The most significant relationship exists between the control over cooling and ventilation (Pearson Coefficient = 0.686; Sig. = 0.000); the least significant relationship exists between the control over lighting and noise (Pearson Coefficient = 0.196; Sig. = 0.000). The number of control opportunities was also significantly related to the degree of control over heating, cooling, ventilation, and lighting, especially to the control over ventilation (Pearson Coefficient = 0.315, Sig. = 0.000) and lighting (Pearson Coefficient = 0.315, Sig. = 0.000). The number of control opportunities was not significantly related to the degree of control over noise. This is because most of the control opportunities listed in this study were about heating, cooling, ventilation, and lighting, while no noise control opportunities were included in this study nor in other studies. Noise control opportunities were not so tangible as an environmental system or interior elements to occupants. Both the number of control opportunities and the degree of control over each aspect except noise were significantly negatively related to adverse perceptions. The most influential aspect is the degree of control over ventilation (Pearson Coefficient = −0.370, Sig. = 0.000).

Table 7. Control opportunities and adverse perceptions.

Control Opportunities (0 = No; 1 = Yes)		Number of Adverse Perceptions		
		Mean	Difference	Significance
Window shades	0	1.4	0.5	0.000
	1	0.9		
Operable window	0	1.4	0.4	0.000
	1	1.0		
Thermostat	0	1.3	0.3	0.012
	1	1.0		
Room air-conditioning unit	0	1.2	0.0	0.969
	1	1.2		
Ceiling fan	0	1.2	0.5	0.087
	1	0.7		
Portable fan	0	1.2	0.0	0.953
	1	1.2		
Adjustable air vent	0	1.2	0.2	0.140
	1	1.0		
Heater	0	1.2	0.5	0.051
	1	0.7		
Door to interior or exterior space	0	1.2	0.1	0.216
	1	1.1		
Light switch	0	1.5	0.5	0.000
	1	1.0		
Light dimmer	0	1.3	0.4	0.017
	1	0.9		
Desk (task) light	0	1.3	0.3	0.013

Table 8. Correlation table.

		Control over Heating	Control over Cooling	Control over Ventilation	Control over Lighting	Control over Noise	Number of Controls	Number of Adverse Perceptions
Control Over Heating	Pearson Correlation	1	0.606 **	0.544 **	0.342 **	0.561 **	0.178 **	−0.111 **
	Sig. (2-tailed)		0.000	0.000	0.000	0.000	0.000	0.010
Control Over Cooling	Pearson Correlation	0.606 **	1	0.686 **	0.628 **	0.270 **	0.290 **	−0.270 **
	Sig. (2-tailed)	0.000		0.000	0.000	0.000	0.000	0.000
Control Over Ventilation	Pearson Correlation	0.544 **	0.686 **	1	0.554 **	0.271 **	0.315 **	−0.370 **
	Sig. (2-tailed)	0.000	0.000		0.000	0.000	0.000	0.000
Control Over Lighting	Pearson Correlation	0.342 **	0.628 **	0.554 **	1	0.196 **	0.315 **	−0.277 **
	Sig. (2-tailed)	0.000	0.000	0.000		0.000	0.000	0.000
Control Over Noise	Pearson Correlation	0.561 **	0.270 **	0.271 **	0.196 **	1	0.070	−0.002
	Sig. (2-tailed)	0.000	0.000	0.000	0.000		0.086	0.965
Number of Controls	Pearson Correlation	0.178 **	0.290 **	0.315 **	0.315 **	0.070	1	−0.142 **
	Sig. (2-tailed)	0.000	0.000	0.000	0.000	0.086		0.000
Number of Adverse Perceptions	Pearson Correlation	−0.111 **	−0.270 **	−0.370 **	−0.277 **	−0.002	−0.142 **	1
	Sig. (2-tailed)	0.010	0.000	0.000	0.000	0.965	0.000	

** Correlation is significant at the 0.01 level (2-tailed).

5. Findings

The study presents a personal control survey in open-plan workplaces. The most common control opportunities available in open-plan offices were light switches, operable windows, and window shades. Respondents had higher degree of control over lighting, cooling, and ventilation than noise and heating. This is because heating is seldom used in Shenzhen which is located in a subtropical climate, and noise control is always a problem in open-plan offices. Figure 1 illustrates the relationships which are statistically significant in this study. Most of control opportunities except desk or task lights were closely associated with the degree of control. Among the 12 control opportunities, only six of them were negatively related to the adverse perceptions. They were window shades, operable windows, thermostat, light switch, light dimmer, and desk (task) light. The degree of control over heating, cooling, ventilation, and lighting were negatively related to the number of adverse perceptions. In sum, some control opportunities supporting the degree of control did not necessarily play a role in alleviating adverse perceptions.

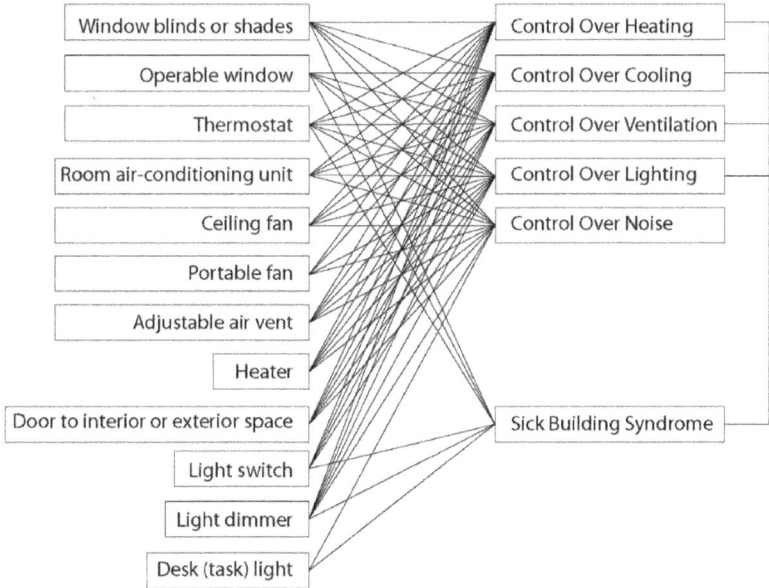

Figure 1. Relationship and effectiveness.

The two research questions raised in the beginning of this study could be responded as follows:

Are the controls (such as operable windows, electric fans, task lights, blinds, and the like.) associated with a high degree of perceived individual control over the physical environment? Most of the control opportunities contributed to significant differences on the control degree perceptions. However, respondents who had desk or task lights available did not perceive that they had significantly more control over heating, cooling, ventilation, and lighting. Among all control opportunities, light dimmers were more likely to appear with other control opportunities while desk/task lights were not likely to do so. The two findings disclosed the conflict between desk or task lights with other control opportunities. In this study, most of workstations with desk or task lights were cubicles with high partitions, where respondents had high privacy but less access to other opportunities. The numbers of control opportunities significantly related to degree of control over heating, cooling, ventilation, and lighting, especially control over ventilation and lighting. The number of control opportunities was not significantly associated with the degree of control over noise.

Are the perceived controls and degree of control over the physical environment negatively associated with adverse perceptions? Respondents who had window blinds or shades, operable windows, thermostats, a light switch, a light dimmer, or desk (task) lights reported significantly fewer adverse perceptions than those who did not have them. On the other side, room air-conditioning units, ceiling fans, portable fans, adjustable air vents, heater, and doors to interior or exterior space did not contribute to a significant difference, which indicates that although these opportunities were important to increasing users' degree of control over the physical environment, they did not play a role in alleviating adverse perceptions. Both the number of control opportunities and the degree of control over the physical environment (except the noise environment) were negatively associated with the number of adverse perceptions. The most influential one was the degree of control over ventilation.

6. Conclusions

Comfort and satisfaction studies address that it is important for occupants to have adaptive opportunities that can help them to alter and control their indoor environmental conditions. This study points out that the mere sum of control opportunities is not a good measure of adaptive opportunities. To assess the usefulness of the personal control in workplaces, the study combined control questions from three systematic occupant survey tools: the U.K. BUS questionnaire, the U.S. IEQ occupant survey, as well as SCATs project in Europe. The results address the following three issues in interior design and the research for effective individual control.

Avoid control opportunities conflicting. Control opportunities should be available for controlling different indoor environments, including heating, cooling, ventilation, lighting, and noise. Some of them are closely related to each other. For example, operable windows and blinds can provide controls over heating, cooling, and lighting. However, some of them may probably conflict with others. For example, private workstations can improve noise environments in open-plan offices while it may reduce accessibility to some controls such as switches or windows which are located outside of the workstation.

Differentiate effective and ineffective control strategies. This study reviewed literature on individual control over the office physical environment, and identified 12 possible control opportunities for occupants working in open-plan workplaces. However, not all of these opportunities are negatively associated with occupants' adverse perceptions. The most effective ones were found out to be operable windows and blinds; while the most ineffective ones were some mechanical solutions (heaters, fans and air-conditioning units), and doors to interior or exterior space. Interior design or environmental control system should consider prioritize non-mechanical personal controls. In office environments, doors should be cautiously used as a control strategy.

Design noise control strategies and interfaces. In this study, the perceived degree of control over noise did not increase as the total number of individual control opportunities increased; meanwhile, the perceived degree of control over noise was not negatively related to adverse perceptions. Most of control opportunities were about heating, cooling, ventilation and lighting. Few studies mentioned noise control, and occupants could hardly think of tangible noise control opportunities. Although previous studies proposed acoustic solutions such as noise mask, how occupants have tangible individual control over noise is still an unanswered question.

Due to the limitations of sampling, the findings and implications could not be generalized to all work environments. There are many factors influencing positive or negative environmental perceptions in workplaces, including demographic, social, and economic ones. Therefore, the association of personal controls with environmental perceptions found in this study should be cautiously interpreted. This study mainly investigated the perception of existence the personal control opportunities and the related effects; while it did not collect the data about how these controls were used, in other words, actual use behaviors of occupants. The future studies should observe the actual use behaviors of occupants on these personal controls to validate this study. How to design effective control opportunities remains an active contentious topic that will need further investigation.

Author Contributions: Z.G., J.Z. and L.S. contributed equally to this paper.

Funding: This research received no external funding.

Acknowledgments: The authors are grateful to all participants for their inputs during this survey.

Conflicts of Interest: The authors declare no conflict of interest.

References

1. Oommen, V.G.; Knowles, M.; Zhao, I. Should health service managers embrace open plan work environments? A review. *Asia Pac. J. Health Manag.* **2008**, *3*, 37–43.
2. Gou, Z.; Lau, S.S.-Y.; Chen, F. Subjective and objective evaluation of the thermal environment in a three-star green office building in china. *Indoor Built Environ.* **2012**, *21*, 412–422. [CrossRef]
3. Hwang, T.; Kim, J.T. Assessment of indoor environmental quality in open-plan offices. *Indoor Built Environ.* **2013**, *22*, 139–156. [CrossRef]
4. Gou, Z.; Lau, S.S.-Y.; Shen, J. Indoor environmental satisfaction in two leed offices and its implications in green interior design. *Indoor Built Environ.* **2012**, *21*, 503–514. [CrossRef]
5. Boje, A. *Open Plan Offices*; Business Books Ltd.: London, UK, 1971.
6. Bosma, H.; Stansfeld, S.A.; Marmot, M.G. Job control, personal characteristics and heart disease. *J. Occup. Health Psychol.* **1998**, *3*, 402–409. [CrossRef] [PubMed]
7. Brill, M.; Keable, E.; Fabiniak, J. The myth of open-plan. *Facil. Des. Manag.* **2000**, *19*, 36–38.
8. Brookes, M.J. Office landscape: Does it work? *Appl. Ergon.* **1972**, *3*, 224–236. [CrossRef]
9. Meijer, E.M.; Frings-Dresen, M.H.W.; Sluiter, J.K. Effects of office innovation on office workers' health and performance. *Ergonomics* **2009**, *52*, 1027–1038. [CrossRef] [PubMed]
10. Gou, Z.; Lau, S.S.-Y. Sick building syndrome in open-plan offices: Workplace design elements and perceived indoor environmental quality. *J. Facil. Manag.* **2012**, *10*, 256–265. [CrossRef]
11. Brager, G.S.; Paliaga, G.; de Dear, R. Operable windows, personal control and occupant comfort. *ASHRAE Trans.* **2004**, *110*, 17–35.
12. Clements-Croome, D. Designing the indoor environment for people. *Archit. Eng. Des. Manag.* **2005**, *1*, 45–55. [CrossRef]
13. Hedge, A.; Khalifa, H.E.; Zhang, J. On the control of envrionmental conditions using personal ventilation systems. In Proceedings of the Human Factors and Ergonomics Society 53rd Annual Meeting, San Antonio, TX, USA, 19–23 October 2009.
14. Vischer, J.C. Towards an environmental psychology of workspace: How people are affected by environments for work. *Archit. Sci. Rev.* **2008**, *51*, 97–108. [CrossRef]
15. Leaman, A.; Bordass, B. Productivity in buildings: The 'killer' variables. In *Creating the Productive Workplace*; Clements-Croome, D., Ed.; Taylor & Francis Group: London, UK; New York, NY, USA, 2006.
16. Loftness, V.; Hartkopf, V. *Building Investment Decision Support (Bids): Cost-Benefit Tool to Promote High Performance Components, Flexible Infrastructures and Systems Integration for Sustainable Commercial Buildings and Prodcutive Organizations*; Center for Building Performance and Diagnostics, Carnegie Mellon University: Pittsburg, PA, USA, 2005.
17. Newsham, G.R.; Aries, M.; Mancini, S.; Faye, G. Individual control of electric lighting in a daylit space. *Light. Res. Technol.* **2008**, *40*, 25–41. [CrossRef]
18. Zeiler, W.; Vissers, D.; Maaijen, R.; Boxem, G. Occupants' behavioural impact on energy consumption: 'Human-in-the-loop' comfort process control. *Archit. Eng. Des. Manag.* **2014**, *10*, 108–130. [CrossRef]
19. Gou, Z. Green building for office interiors: Challenges and opportunities. *Facilities* **2016**, *34*, 614–629. [CrossRef]
20. Vischer, J.C. *Space Meets Status: Designing Workplace Performance*; Routledge: Oxon, UK, 2005.
21. Liu, J.; Yao, R.; Wang, J.; Li, B. Occupants' behavioural adaptation in workplaces with non-central heating and cooling systems. *Appl. Therm. Eng.* **2012**, *35*, 40–54. [CrossRef]
22. De Dear, R.; Brager, G.; Cooper, D. *Developing an Adaptive Model of Thermal Comfort and Preference*; Final Report; American Society of Heating, Refrigerating and Air-conditioning Engineers, Inc.: Atlanta, GA, USA, 1997.
23. Andersen, R.V. Occupant Behaviour with Regard to Control of the Indoor Environment. Ph.D. Thesis, Technical University of Denmark, Lyngby, Denmark, 2009.

Buildings **2018**, *8*, 110

24. Paciuk, M. The Role of Personal Control of the Environment in Thermal Comfort and Satisfaction at the Workplace. Ph.D. Thesis, The University of Wisconsin-Milwaukee, Milwaukee, WI, USA, 1989.

25. Nicol, F.; Roaf, S. Post-occupancy evaluation and field studies of thermal comfort. *Build. Res. Inf.* **2005**, *33*, 338–346. [CrossRef]

26. Huizenga, C.; Abbaszadeh, S.; Zagreus, L.; Arens, E. Air quality and thermal comfort in office buildings: Results of a large indoor environmental quality survey. In Proceedings of the 8th International Conference and Exhibition on Healthy Buildings, Lisbon, Portugal, 4–8 June 2006.

 © 2018 by the authors. Licensee MDPI, Basel, Switzerland. This article is an open access article distributed under the terms and conditions of the Creative Commons Attribution (CC BY) license (http://creativecommons.org/licenses/by/4.0/).

Article

Are Mental Biases Responsible for the Perceived Comfort Advantage in "Green" Buildings?

Mattias Holmgren * and Patrik Sörqvist

Department of Building, Energy and Environmental Engineering, University of Gävle, SE-801 76 Gävle, Sweden; patrik.sorqvist@hig.se
* Correspondence: mattias.holmgren@hig.se; Tel.: +46-73-407-3817

Received: 8 January 2018; Accepted: 27 January 2018; Published: 30 January 2018

Abstract: Previous research has shown that merely calling an indoor environment environmentally certified will make people favor that environment over a conventional alternative. In this paper we explore whether this effect depends on participants deliberately comparing the two environments, and whether different reasons behind the certification influence the magnitude of the effect. In Experiment 1, participants in a between-subjects design assigned higher comfort ratings to an indoor environment that had been labeled "environmentally certified" in comparison with the exact same indoor environment that was unlabeled, suggesting that the effect arises even when participants do not compare the two environments when making their estimates. The results from Experiment 2 indicate that climate change mitigation (as the reason for the certification) is a slightly better trigger of the effect compared to climate change adaptation. The results suggest that studies on psychological effects of "green" buildings should experimentally control for the influence from participants' judgmental biases.

Keywords: eco-label effect; bias; comfort; environmental certification; "green" buildings

---

1. Introduction

Buildings consume roughly 40% of global energy and influence the environment substantially [1]. Households in the European Union are responsible for 20% of the total GHG emissions [2] and for up to 26% of the total energy consumption [3]. Because of this, the European Parliament have implemented Directive 2010/31/EU, which states that all new buildings—from the 1st of January 2021—in the European Union should incorporate energy saving measures and be "nearly zero energy buildings" [4]. An important step in this endeavor is the environmental certification of buildings [5]. Environmentally certified (or "green") buildings are better for the environment [6]. A perhaps more surprising effect of environmental certification is that "green" buildings also seem to be better for the inhabitants [7,8], as occupants are more satisfied with the indoor environment of "green" buildings compared to a conventional counterpart [9,10]. In addition, one study conducted by Holmgren, Kabanshi, and Sörqvist [11] showed that it is enough to call an indoor environment environmentally certified to make people favor that environment over a conventional alternative, even when the two alternatives are actually identical. In the current paper, we further explore the notion of whether people's biases contribute to the psychological benefits of "green" buildings.

1.1. "Green" Buildings and the Psychological Effects of Environmental Certification

Several countries have developed their own "green" building assessment tools. Examples include The Building Research Establishment Environmental Assessment Method (U.K.), Hong Kong Building Environmental Assessment Methods, Leadership in Energy and Environmental Design (U.S.), Green Star (Australia), and Green Building Label (China). "Green" buildings can be described as "healthy facilities designed and built in a resource-efficient manner, using ecologically based principles" [12], and they are, compared to conventional buildings, indeed better for the environment, as they preserve natural resources [13,14], protect the eco-system [15], mitigate environmental hazards [13,16], and improve energy efficiency [14,17]. They are also (in some cases) better than conventional buildings in other capacities such as economic [18] and ergonomic dimensions [19,20].

Furthermore, implementation of energy-efficient measures in buildings can lead to changes in the indoor environment [21] and, as stated above, in some cases, improvements in the indoor climate for human occupants [19]. Making a building "green" seems to improve occupants' subjective evaluation of the indoor environment [22], as well as their performance on cognitive tasks [23]. Inhabitants in energy-efficient houses perceive their homes as comfortable, both thermally [24,25] and acoustically [25]. Moreover, employees in "green" buildings and offices have a high overall workplace satisfaction [7], and occupants are more satisfied with the indoor environment of "green" buildings than the indoor environment of a conventional counterpart [9,10,26,27]. It is also important to note, however, that there are studies that show no significant differences between "green" and non-"green" buildings (e.g., [28]). Moreover, "green" building occupants are more forgiving of their buildings (e.g., balancing the good features against the bad; [29]). This tolerance for bad building features seems to be related to people's degree of environmental concern [30]. There are also studies suggesting that "green" buildings have the ability to improve work productivity [8,31]. There is, however, research that challenges the claims regarding improvements of productivity [32], which might be due to certain biases in self-evaluation measures (e.g., questionnaires; [33]).

Collectively, the studies of the psychological benefits of "green" buildings are promising, but it is yet unclear how the participants' expectations and stereotypical beliefs about "green" buildings bias these self-reported subjective evaluations of the indoor environments. Environmental psychology research shows that it is enough to label a consumable product [34] or an artifact such as a desktop lamp [35] "environmentally friendly" to make people favor this alternative over another alternative labeled "conventional", even when the two alternatives are actually identical. For example, people prefer the taste of a cup of coffee labeled eco-friendly to a cup labeled conventional, even when the two cups contain the exact same coffee [34]. This eco-label effect is a robust phenomenon and has been replicated many times (e.g., [11,36–39]). The eco-label effect is a specific instance of a phenomenon called the "placebo effect" [40], whereby an effect arises from a manipulation because people believe the manipulation has an effect, not because the manipulation actually has an effect (see also "framing effects"; [41]).

An eco-label effect has also been shown for labeling of the built environment. For example, people assign higher comfort ratings to an indoor environment called environmentally certified compared to an indoor environment called conventional, even when the two environments are, in fact, identical [11]. Hence, it may well be that the "green" building label attached to environmentally certified buildings also triggers expectations and preconceptions that bias the psychological assessment of the buildings. The vast majority of studies measuring psychological effects of "green" buildings (e.g., [7,22,24,25]) have, to the best of our knowledge, not had experimental control over the potential influence of these biases. Because of this, the reason why people report psychological benefits from their experiences with "green" buildings could be, at least in part, because they believe in these benefits rather than actual benefits.

1.2. Purpose

Drawing on the work of Holmgren et al. [11], the purpose of this study was to further investigate whether the subjective evaluation of an indoor environment is biased such that people say they prefer the indoor environment of a building they believe is environmentally certified. Specifically, we explored whether this effect arises using a between participants design (Experiment 1). Most (if not all) previous studies on the eco-label effect have employed a within-participants design wherein the participants make estimates of both the "eco-labeled" and the "conventional" alternative. This procedure encourages the participants to compare the two alternatives when making the estimates. The point with using a between-subjects design, wherein the participants only make estimates of either the "eco-labeled" or the "conventional" alternative, is that this comparison is not reinforced by the experimental procedure. The present series of experiments also explored whether different reasons behind the certification influence the magnitude of the effect (Experiment 2).

2. Experiment 1

The purpose of Experiment 1 was to test whether the label effect arises in a between-subjects design. The participants were allocated to one of two conditions: half of the participants were assigned to a framing condition in which the participants were told that the building they were in was environmentally certified (hereinafter referred to as the *framed condition*), and the other half was assigned to a control condition, which had no framing (hereinafter referred to as the *control condition*). We hypothesized that there would be an effect of labeling the building "environmentally certified" on perceived overall comfortableness. More specifically, we hypothesized that the participants in the framed condition would rate the indoor environment as overall more comfortable compared to the participants in the control condition, even though the two participant groups evaluated the exact same indoor environment.

3. Method—Experiment 1

3.1. Participants

A total of 42 students recruited at the University of Gävle (64% women) participated in the experiment (mean age = 24.67, SD = 4.22). They all received a small honorarium for their participation. The study was approved by the Research Ethics Review Board at Uppsala University (Dnr 2015/475). Oral consent was considered to be sufficient by the ethics review board. The data collectors took note of the oral consent.

3.2. Materials

A questionnaire was used to obtain data. On the first page of the questionnaire, all participants (in both conditions) were told that the University had implemented a survey regarding how students and personnel perceive the University's premises in terms of comfortableness. For half of the questionnaires (i.e., the "framed condition"), the following statement was also added to the introductory text: "*The laboratory, where you are now, is environmentally certified according to ISO 14001 and is provided with environmentally friendly electricity, ventilation and heating. The vision of the University of Gävle is to strive toward a sustainable growth within all its practices, and the University's business should be conducted in such a way that the positive impact on the existing environment increases and the negative impact decreases*". For the other half of the questionnaires (i.e., the control condition), no such information was provided. After reading the introductory texts, the participants rated overall perceived comfortableness in the room in which they were presently sitting, on a scale ranging from 1 (i.e., not at all comfortable) to 7 (i.e., very comfortable). In the "control condition", the question was: "*How comfortable would you say the room is, generally?*" In the "framed condition", the question to respond to was: "*How comfortable would you say the **environmentally certified** room is, generally?*" (The bold text was added to highlight the difference).

3.3. Design and Procedure

A between-participants design was used with framing as the independent variable with two levels: framing of the indoor environment as "environmentally certified" versus a no-framing control condition. The participants were randomly distributed across the two conditions. This randomization resulted in a sample of 12 women and 9 men in the framing condition and 15 women and 6 men in the control condition (the two conditions were highly matched with regard to participants' age).

The experiment took place in a laboratory at the University at Gävle and was part of a larger data collection on research questions without relation to the current research questions. Two identical rooms were used, and the two rooms were used an equal number of times in both conditions. The participants sat alone in the laboratory room when making the estimate and responded to the questionnaire. Each participant had spent approximately 45 min in the room before responding to the questionnaire.

4. Results and Discussion—Experiment 1

The participants in the framed condition perceived the overall comfortableness of the room as greater ($M = 5.62$, $SD = 0.92$) than the participants in the control condition did ($M = 4.52$, $SD = 1.44$). This result was statistically significant, as shown with the one-way analysis of variance across the two conditions, $F(1, 40) = 8.66$, $p = 0.005$, $\eta_p^2 = 0.18$. Hence, the results from Experiment 1 suggest that self-reported estimates of indoor environments are more favorable when the person making the estimates thinks (or knows) that the building is environmentally certified. This finding replicated the effect showed by Holmgren et al. [11] and showed that the effect can arise in the context of a between-participants design wherein the participants do not compare the two environments, in contrast to a within-participants design.

5. Experiment 2

The purpose of Experiment 2 was to replicate the main finding from Experiment 1 and to explore whether the magnitude of the label effect depends on what the participants think is the reason behind the "environmental certification". To cope with global warming, scientists and policymakers have pointed at the importance of interventions that *adapt to* the climate change and *mitigate* the climate change [42–44]. Social interventions that can be used to mitigate (slow down) climate change are, for example, promotion of green consumption [45] and promotion of recycling [46]. The climate change discourse has, over the last decades, had its focus on climate change mitigation. However, as the effects of climate change already are occurring [42], it is also important to consider possibilities to adapt to climate change, for example by implementing new technological solutions that can cope with the impacts of global warming. As implementations of both of these environmental policies—mitigation and adaptation—are essential to cope with climate change, it would be interesting to test whether people's beliefs about climate change mitigation and adaptation (as different reasons for the environmental certification) influence the magnitude of the "green" label effect on perceived comfort. To test this, some participants in Experiment 2 were told that the "environmental certification" was a result of mitigation interventions (measures taken to slow down climate change), while others were told that it was a result of adaptation interventions (measures taken to adapt the building to climate change). As climate change mitigation is more familiar to the public than climate change adaptation, and because adaptation has had a bad reputation in the past [47,48], we hypothesized that the participants in the mitigation condition would be more susceptible to the "green" label effect on perceived overall comfortableness compared to the participants in the adaptation condition.

6. Method—Experiment 2

6.1. Participants

A total of 135 students recruited at the University of Gävle (49% women) participated in the experiment (mean age = 25.29, $SD = 6.21$). Students were invited to take part in the study when they had arrived to their classroom prior to lectures. The study was approved by the Research Ethics Review Board at Uppsala University (Dnr 2015/475). Oral consent was considered to be sufficient by the ethics review board. The data collectors took note of the oral consent.

6.2. Materials

The questionnaires were identical to those in Experiment 1, except for the information provided prior to the comfort rating. One third of the questionnaires framed the environmental certification as a result of mitigation interventions: *"The University of Gävle's vision is to **mitigate** the current climate change through the built environment. Examples of how the University of Gävle **mitigates climate change** are: installation of solar films on the windows and painting the roofs in a lighter color to alleviate the load on the air conditioning, and installation of solar panels that supply the University with natural energy. These **mitigation actions** have made the room where you are now are sitting environmentally certified according to ISO 14001."* One third of the questionnaires framed the environmental certification as a result of adaptation interventions: *"The University of Gävle's vision is to **adapt** the built environment to the current climate change. Examples of how the University of Gävle **adapts to climate change** are: installation of solar films on the windows and painting the roofs in a lighter color to alleviate the load on the air conditioning, and installation of solar panels that supply the University with natural energy. These **adaptation actions** have made the room where you now are sitting environmentally certified according to ISO 14001."* The questionnaire distributed in the control condition was identical to the one in the control condition of Experiment 1.

6.3. Design and Procedure

A between-participants design was used with framing as the independent variable with three levels: two "environmentally certified" framing conditions, in which the participants were told that the certification was either a result of mitigation ($N = 46$) or of adaptation interventions ($N = 46$), and a control condition ($N = 43$). The experiment took place at the University at Gävle. The participants were randomly distributed across conditions. This randomization resulted in 21 women and 25 men in the adaptation condition, 28 women and 18 men in the mitigation condition, and 17 women and 26 men in the control condition (age distribution was highly matched between conditions). The participants sat in a classroom, and they were invited to take part in a survey on perceived comfortableness in classrooms. The three different questionnaires were distributed among the students in the classroom who were willing to participate. Participants next to each other received the same type of questionnaire to prevent them from noticing that other participants had received a questionnaire that was different from the one they had received themselves. The data collection was repeated a number of times in different classrooms with different classes. The participants had spent approximately 15 min in the classroom before making the evaluation. The three questionnaires were equally distributed within each class and classroom.

7. Results and Discussion—Experiment 2

As can be seen in Figure 1, the participants perceived the classroom as more comfortable in the mitigation condition compared to the control condition, whereas there was only a tendency for perceiving the room as more comfortable in the adaptation condition compared to the control condition. There was no difference between the mitigation and the adaptation condition. A one-way analysis of variance across the three conditions was calculated with overall comfortableness ratings as the dependent variable. The analysis revealed a significant difference between conditions, $F(2, 132) = 3.51$, $p = 0.033$, $\eta_p^2 = 0.05$. Independent samples t-tests revealed that the participants in the mitigation condition ($M = 4.91$, $SD = 0.96$) perceived the classroom as more comfortable than did the participants in the control condition ($M = 4.42$, $SD = 0.73$), $t(87) = 2.72$, $p = 0.008$, $\eta^2 = 0.08$. The classroom was also perceived as more comfortable in the adaptation condition ($M = 4.77$, $SD = 0.99$) compared to the control condition ($M = 4.42$, $SD = 0.73$), but this difference was not statistically significant with the conventional alpha threshold, $t(87) = 1.91$, $p = 0.059$, $\eta^2 = 0.04$. No difference was found between the mitigation and the adaptation condition, $t(90) = 0.69$, $p = 0.490$, $\eta^2 = 0.01$.

Experiment 2 revealed a framing effect on perceived overall comfortableness, congruent with Experiment 1 and previous research [11]. The framing effect appeared somewhat stronger in the mitigation condition, but there was no difference between the two framing conditions. Because of this, the safest conclusion from Experiment 2 is arguably that the psychological evaluation of environmentally certified buildings is biased, but the reason for this environmental certification—mitigation or adaptation—has only marginal effect.

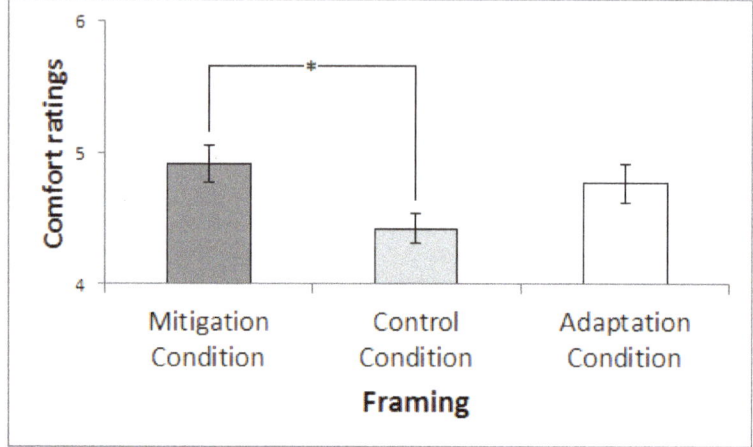

Figure 1. Mean perceived overall comfort ratings assigned to a classroom in a building. The participants were either told that a mitigation intervention or that an adaptation intervention had led to the building being environmentally certified or they were not at all told that the building was environmentally certified. Error bars represent standard error of means. Note: * Significant at alpha = 0.05.

8. General Discussion

In Experiment 1, participants assigned higher comfort ratings to an indoor environment labeled "environmentally certified" in comparison with the exact same but unlabeled indoor environment. Experiment 2 replicated the results from Experiment 1 and found that the magnitude of this preference bias for environmentally certified buildings is quite insensitive to the reasons for the certification. If anything, the strongest effect was found when the participants were told that the reason for the environmental certification was climate-change mitigation.

A large body of studies have measured potential benefits to be gained from working or living in "green" buildings by obtaining self-reports from participants [7–10,22,26,27,49]. As shown here, these self-reports are sensitive to framing effects and may therefore be unreliable as a measure of the "green" building benefits for the inhabitants. What seem to be psychological benefits from engineering interventions of buildings may reflect the consequences of people's biases and beliefs about environmental certification rather than effects of physical differences between the built environments.

It should be stated that our intention here was not to conclude that all benefits for occupants in "green" buildings simply are a consequence of the occupants' biases. Allen et al. [23] showed, for example, that occupants obtain higher scores on cognitive tasks when the tasks are conducted in "green" buildings compared to in conventional buildings, even when the participants were blinded to what condition they were in (i.e., the participants did not know whether they conducted the task in an indoor environment of a "green" or a conventional building). We rather conclude from the present study that people's biases influence the magnitude of the effects associated with "green" buildings, something that calls for the need to control for this issue in scientific endeavors in the future.

It is important to stress that even though a placebo-like effect improves people's perception of the indoor environment of "green" buildings [11], due to their expectations and biases, the feelings of greater comfort in "green" buildings may still be "real". In other words, people are more comfortable in environmentally certified buildings, even though the effect may surface through their expectations rather than through a physical difference between "green" and "conventional" built environments. It is possible, for example, that the perception of a room as more comfortable, because of a "green" building label, is beneficial for learning abilities and health, similar to actual high-quality indoor environments [50,51].

A step towards understanding how to enhance the psychological benefits of environmental certification was taken in Experiment 2. This experiment indicated that a mitigation frame could potentially be an efficient trigger of the label effect. A possible explanation for the slightly stronger effect of the mitigation framing, in comparison with the adaptation framing, is that adaptation to climate change has had poor reputation [47,48] compared to mitigation. Thus, participants might have felt slightly more indifferent toward the framing information regarding the University's vision to adapt to climate change when told this was the reason for the certification. Additionally, as the climate change discourse over the last decades has had its focus on climate change mitigation, adaptation could have been a new concept to the participants and was perhaps therefore not an efficient trigger. Furthermore, environmental concern [11,35] and attitudes toward behaviors that mitigate climate change (e.g., pro-environment consumer behavior; [34,37]) co-vary with the magnitude of the eco-label effect (e.g., the tendency to report that the lighting from a light source is more comfortable when the light source is labeled "environmentally friendly" compared to when the same light source is labeled "conventional"). Additionally, it is probable that the mitigation frame is more appealing to people with high environmental concern than the adaptation frame, making the effect of the mitigation frame somewhat stronger. It should be noted, though, that no significant difference between the mitigation and the adaptation frame conditions was obtained in the present study.

On a methodological note, it is worth mentioning that previous studies on the eco-label effect have consistently used a within-participants design, in which the same persons evaluated both an "eco-labeled" and a "conventional" (or non-labeled) target (e.g., [11,35,37,38]). This procedure, in which the same person is making estimates of two products, encourages a comparison between the "eco-labeled" and the "conventional" alternative, and this encouragement may exaggerate the differences between the two alternatives in the self-reported estimates. Conversely, the present study used a between-participants design in which the participants who were asked to evaluate the "environmentally certified" environment were not the same as those who evaluated the non-labeled environment. As shown in the current study, the eco-label effect is robust and reliable enough to arise even in a between-participants design in which the participants do not compare the two alternatives. This finding may also be of practical importance, because this robustness of the effect shows that the psychological framing effects of environmental certification may appear even when people are not comparing certified and non-certified buildings.

8.1. Limitations of the Study and Directions for Future Research

The current study is (to our knowledge) the first to show the "green" label effect in the context of a between-participants design. The study has, however, some limitations: (1) the sample sizes were quite small (especially in Experiment 1), (2) the participants were university students, and (3) the settings were either two rooms in a laboratory or classrooms. Studies regarding occupant satisfaction in "green" buildings are usually conducted in office buildings, with office workers as respondents and with larger sample sizes compared to what was used in the present study (e.g., [52–54]). Even so, past research has also looked at comfort in "green" university buildings [49,55].

Furthermore, there was no control for the potential influence of clothing and metabolic rate in the current study. These factors can influence perceived comfort [56]. Another limitation of the current study is that the only dependent measure was overall comfort. Comfort in the indoor environment is determined by sub-variables; for example, air quality [57], thermal [58], acoustic [59], and visual comfort [60]. These variables were not treated individually in the current study. Furthermore, environmental concern is related to acceptance for an unpleasant indoor environment in "green" buildings [30], and people with high environmental concern are more susceptible to the eco-label effect (e.g., [11,34]). Given the information above, to achieve a more comprehensive understanding of how the environmental certification label influences indoor environment perception, and to investigate how generalizable this effect is, future research should control for the previously mentioned variables, include additional dependent variables, and consider replicating this study with larger sample sizes, different demographic variables, and different settings. Targets for future research also include investigation into whether reasons for environmental certification, other than mitigation and adaptation to climate change, can influence the magnitude of the effect.

8.2. Conclusions

The bias toward a preference for environmentally certified buildings appears to be small but reliable and appears across different environmental settings and experimental setups. This conclusion stresses the need to control for this bias when investigating the psychological benefits of "green" buildings.

Author Contributions: M.H. and P.S. contributed to the study design. Testing and data-collection was performed by M.H. M.H. performed the data analyses under the supervision of P.S. M.H. and P.S. drafted the manuscript. Both authors approved the final version of the manuscript.

Conflicts of Interest: The authors declare no conflict of interest.

References

1. GhaffarianHoseini, A.; Dahlan, N.D.; Berardi, U.; GhaffarianHoseini, A.; Makaremi, N.; GhaffarianHoseini, M. Sustainable energy performances of green buildings: A review of current theories, implementations and challenges. *Renew. Sustain. Energy Rev.* **2013**, *25*, 1–17. [CrossRef]
2. European Environment Agency. Final Energy Consumption by Sector. 2013. Available online: http://www.eea.europa.eu/data-and-maps/indicators/final-energyconsumption-by-sector-5/assessment#toc-1 (accessed on 15 June 2015).
3. Pérez-Lombard, L.; Ortiz, J.; Pout, C. A review on buildings energy consumption information. *Energy Build.* **2008**, *40*, 394–398. [CrossRef]
4. European Union. EU directive 2010/31 of the European parliament and of the council of 19 May on the energy performance of buildings. *Off. J. Eur. Union* **2010**, *2010*, L153.
5. Zuo, J.; Zhao, Z.Y. Green building research—Current status and future agenda: A review. *Renew. Sustain. Energy Rev.* **2014**, *30*, 271–281. [CrossRef]
6. Chwieduk, D. Towards sustainable-energy buildings. *Appl. Energy* **2003**, *76*, 211–217. [CrossRef]
7. Kato, H.; Too, L.; Rask, A. Occupier perceptions of green workplace environment: The Australian experience. *J. Corp. Real Estate* **2009**, *11*, 183–195. [CrossRef]
8. Thatcher, A.; Milner, K. Changes in productivity, psychological wellbeing and physical wellbeing from working in a 'green' building. *Work* **2014**, *49*, 381–393. [PubMed]
9. Leder, S.; Newsham, G.R.; Veitch, J.A.; Mancini, S.; Charles, K.E. Effects of office environment on employee satisfaction: A new analysis. *Build. Res. Inf.* **2015**, *44*, 34–50. [CrossRef]
10. Holopainen, R.; Salmi, K.; Kähkönen, E.; Pasanen, P.; Reijula, K. Primary energy performance and perceived indoor environment quality in Finnish low-energy and conventional houses. *Build. Environ.* **2015**, *87*, 92–101. [CrossRef]
11. Holmgren, M.; Kabanshi, A.; Sörqvist, P. Occupant perception of "green" buildings: Distinguishing physical and psychological factors. *Build. Environ.* **2017**, *114*, 140–147. [CrossRef]
12. Kibert, C.J. *Sustainable Construction: Green Building Design and Delivery*; John Wiley and Sons, Inc.: Hoboken, NJ, USA, 2008.
13. Coelho, A.; de Brito, J. Influence of construction and demolition waste management on the environmental impact of buildings. *Waste Manag.* **2012**, *32*, 532–541. [CrossRef] [PubMed]
14. Thormark, C. The effect of material choice on the total energy need and recycling potential of a building. *Build. Environ.* **2006**, *41*, 1019–1026. [CrossRef]
15. Bianchini, F.; Hewage, K. How green are the green roofs? Lifecycle analysis of green roof materials. *Build. Environ.* **2012**, *48*, 57–65. [CrossRef]

16. Wang, W.; Zmeureanu, R.; Rivard, H. Applying multi-objective genetic algorithms in green building design optimization. *Build. Environ.* **2005**, *40*, 1512–1525. [CrossRef]
17. Turner, C.; Frankel, M. *Energy Performance of LEED for New Construction Buildings*; New Buildings Institute: Vancouver, WA, USA, 2008.
18. Lau, L.C.; Tan, K.T.; Lee, K.T.; Mohamed, A.R. A comparative study on the energy policies in Japan and Malaysia in fulfilling their nations' obligations towards the Kyoto Protocol. *Energy Policy* **2009**, *37*, 4771–4778. [CrossRef]
19. Liu, L.; Rohdin, P.; Moshfegh, B. Evaluating indoor environment of a retrofitted multi-family building with improved energy performance in Sweden. *Energy Build.* **2015**, *102*, 32–44. [CrossRef]
20. Zhang, Y.; Altan, H. A comparison of the occupant comfort in a conventional high-rise office block and a contemporary environmentally-concerned building. *Build. Environ.* **2011**, *46*, 535–545. [CrossRef]
21. Karlsson, J.F.; Moshfegh, B. Energy demand and indoor climate in a low energy building—Changed control strategies and boundary condition. *Energy Build.* **2006**, *38*, 315–326. [CrossRef]
22. Armitage, L.; Murugan, A.; Kato, H. Green offices in Australia: A user perception survey. *J. Corp. Real Estate* **2011**, *13*, 169–180. [CrossRef]
23. Allen, J.G.; MacNaughton, P.; Satish, U.; Santanam, S.; Vallerino, J.; Spengler, J.D. Associations of cognitive function scores with carbon dioxide, ventilation, and volatile organic compound exposures in office workers: A controlled exposure study of green and conventional office environments. *Environ. Health Perspect.* **2015**. [CrossRef] [PubMed]
24. Derbez, M.; Berthineau, B.; Cochet, V.; Lethrosne, M.; Pignon, C.; Riberon, J.; Kirchner, S. Indoor air quality and comfort in seven newly built, energy-efficient houses in France. *Build. Environ.* **2014**, *72*, 173–187. [CrossRef]
25. Derbez, M.; Berthineau, B.; Cochet, V.; Pignon, C.; Riberon, J.; Wyart, G.; Mandin, C.; Kirchner, S. A 3-year follow-up of indoor air quality and comfort in two energy-efficient houses. *Build. Environ.* **2014**, *72*, 288–299. [CrossRef]
26. Liang, H.; Chen, C.; Hwang, R.; Shih, W.; Lo, S.; Liao, H. Satisfaction of occupants toward indoor environment quality of certified green office buildings in Taiwan. *Build. Environ.* **2014**, *72*, 232–242. [CrossRef]
27. Pei, Z.; Lin, B.; Liu, Y.; Zhu, Y. Comparative study on the indoor environment quality of green office buildings in China with a long-term field measurement and investigation. *Build. Environ.* **2015**, *84*, 80–88. [CrossRef]
28. Gou, Z.; Lau, S.S.Y.; Shen, J. Indoor environmental satisfaction in two LEED offices and its implications in green interior design. *Indoor Built Environ.* **2012**, *21*, 503–514. [CrossRef]
29. Gou, Z.; Prasad, D.; Lau, S.S.Y. Are green buildings more satisfactory and comfortable? *Habitat Int.* **2013**, *39*, 156–161. [CrossRef]
30. Deuble, M.P.; de Dear, R.J. Green occupants for green buildings: The missing link? *Build. Environ.* **2012**, *56*, 21–27. [CrossRef]
31. Fisk, W.J. Health and productivity gains from better indoor environments and their relationship with building energy efficiency. *Annu. Rev. Energy Environ.* **2000**, *25*, 537–566. [CrossRef]
32. Byrd, H.; Rasheed, E.O. The productivity paradox in green buildings. *Sustainability* **2016**, *8*, 347. [CrossRef]
33. Rasheed, E.O.; Byrd, H. Can self-evaluation measure the effect of IEQ on productivity? A review of literature. *Facilities* **2017**, *35*, 601–621. [CrossRef]
34. Sörqvist, P.; Hedblom, D.; Holmgren, M.; Haga, A.; Langeborg, L.; Nöstl, A.; Kågström, J. Who needs cream and sugar when there is eco-labeling? Taste and willingness to pay for "eco-friendly" coffee. *PLoS ONE* **2013**, *8*, e80719. [CrossRef] [PubMed]
35. Sörqvist, P.; Haga, A.; Holmgren, M.; Hansla, A. An eco-label effect in the built environment: Performance and comfort effects of labeling a light source environmentally friendly. *J. Environ. Psychol.* **2015**, *42*, 123–127. [CrossRef]
36. Ekelund, L.; Fernqvist, F.; Tjärnebo, H. Consumer preferences for domestic and organically labelled vegetables in Sweden. *Acta Agric. Scand. Sect. C Food Econ.* **2007**, *4*, 229–236. [CrossRef]
37. Lee, W.J.; Shimizu, M.; Kniffin, K.M.; Wansink, B. You taste what you see: do organic labels bias taste perception? *Food Qual. Preference* **2013**, *29*, 33–39. [CrossRef]
38. Sörqvist, P.; Haga, A.; Langeborg, L.; Holmgren, M.; Wallinder, M.; Nöstl, A.; Seager, P.B.; Marsh, J.E. The green halo: Mechanisms and limits of the eco-label effect. *Food Qual. Preference* **2015**, *43*, 1–9. [CrossRef]

39. Wiedmann, K.P.; Hennigs, N.; Behrens, S.H.; Klarmann, C. Tasting green: An experimental design for investigating consumer perception of organic wine. *Br. Food J.* **2014**, *116*, 197–211. [CrossRef]

40. Price, D.D.; Finniss, D.G.; Benedetti, F. A comprehensive review of the placebo effect: Recent advances and current thought. *Annu. Rev. Psychol.* **2008**, *59*, 565–590. [CrossRef] [PubMed]

41. Allison, R.I.; Uhl, K.P. Influence of beer brand identification on taste perception. *J. Mark. Res.* **1964**, *1*, 36–39. [CrossRef]

42. Intergovernmental Panel on Climate Change. *Climate Change 2007: The Physical Science Basis, Contribution of Working Group I to the Fourth Assessment Report of the Intergovernmental Panel on Climate Change*; Solomon, S., Ed.; Cambridge University Press: Cambridge, UK, 2007.

43. Keskitalo, E.C.H. *Climate Change and Globalization in the Arctic: An Integrated Approach to Vulnerability Assessment*; Earthscan: London, UK, 2012.

44. National Research Council. Adapting to the impacts of climate change. In *International Journal of Health Geographics*; National Research Council: Washington, DC, USA, 2010; Volume 7.

45. Heiskanen, E.; Pantzar, M. Toward sustainable consumption: Two new perspectives. *J. Consum. Policy* **1997**, *40*, 409–442. [CrossRef]

46. Menikpura, S.N.M.; Gheewala, S.H.; Bonnet, S.; Chiemchaisri, C. Evaluation of the effect of recycling on sustainability of municipal solid waste management in Thailand. *Waste Biomass Valor* **2013**, *4*, 237–257. [CrossRef]

47. Pielke, R.; Prins, G.; Rayner, S.; Sarewitz, D. Climate change 2007: Lifting the taboo on adaptation. *Nature* **2007**, *445*, 597–598. [CrossRef] [PubMed]

48. Victor, D.; Kennell, C.F.; Ramanathan, V. The climate threat we can beat. *Foreign Aff.* **2012**, *119*, 112–121.

49. Bonde, M.; Ramirez, J. A post-occupancy evaluation of a green rated and conventional on-campus residence hall. *Int. J. Sustain. Built Environ.* **2015**, *4*, 400–408. [CrossRef]

50. Coley, D.A.; Greeves, R.; Saxby, B.K. The effect of low ventilation rates on the cognitive function of a primary school class. *Int. J. Vent.* **2007**, *6*, 107–112. [CrossRef]

51. Rosenfeld, S. Worker productivity: Hidden HVAC cost. *J. Heat. Vent. Air Cond.* **1990**, *9*, 69–70.

52. Abbaszadeh, S.; Zagreus, L.; Lehrer, D.; Huizenga, C. Occupant Satisfaction with Indoor Environmental Quality in Green Buildings. In Proceedings of the Eighth International Conference for Healthy Buildings, 2006: Creating a Healthy Indoor Environment for People, Lisbon, Portugal, 4–8 June 2006.

53. Altomonte, S.; Schiavon, S. Occupant satisfaction in LEED and non-LEED certified buildings. *Build. Environ.* **2013**, *68*, 66–76. [CrossRef]

54. Brown, Z.; Cole, R.J. Influence of occupants' knowledge on comfort expectations and behaviour. *Build. Res. Inf.* **2009**, *37*, 227–245. [CrossRef]

55. Hedge, A.; Miller, L.; Dorsey, J.A. Occupant comfort and health in green and conventional university buildings. *Work* **2014**, *49*, 363–372. [PubMed]

56. Havenith, G.; Holmér, I.; Parsons, K. Personal factors in thermal comfort assessment: Clothing properties and metabolic heat production. *Energy Build.* **2002**, *34*, 581–591. [CrossRef]

57. Frontczak, M.; Schiavon, S.; Goins, J.; Arens, E.; Zhang, H.; Wargocki, P. Quantitative relationships between occupant satisfaction and satisfaction aspects of indoor environmental quality and building design. *Indoor Air* **2012**, *22*, 119–131. [CrossRef] [PubMed]

58. Frontczak, M.; Wargocki, P. Literature survey on how different factors influence human comfort in indoor environments. *Build. Environ.* **2011**, *46*, 922–937. [CrossRef]

59. Navai, M.; Veitch, J.A. *Acoustic Satisfaction in Open-Plan Offices: Review and Recommendations*; Institute for Research in Construction: Ottawa, ON, Canada, 2003.

60. Galasiu, A.D.; Veitch, J.A. Occupant preferences and satisfaction with the luminous environment and control systems in daylit offices: A literature review. *Energy Build.* **2006**, *38*, 728–742. [CrossRef]

© 2018 by the authors. Licensee MDPI, Basel, Switzerland. This article is an open access article distributed under the terms and conditions of the Creative Commons Attribution (CC BY) license (http://creativecommons.org/licenses/by/4.0/).

Article

Design Lessons from Three Australian Dementia Support Facilities

Hing-wah Chau [1],*, Clare Newton [1], Catherine Mei Min Woo [1], Nan Ma [1], Jiayi Wang [1] and Lu Aye [2]

[1]   Faculty of Architecture, Building and Planning, The University of Melbourne, Melbourne,
       VIC 3010, Australia; c.newton@unimelb.edu.au (C.N.); catherine.woo@unimelb.edu.au (C.M.M.W.);
       nan.ma@unimelb.edu.au (N.M.); j.wang244@student.unimelb.edu.au (J.W.)
[2]   Renewable Energy and Energy Efficiency Group, Department of Infrastructure Engineering, Melbourne
       School of Engineering, The University of Melbourne, Melbourne, VIC 3010, Australia; lua@unimelb.edu.au
*   Correspondence: chauh@unimelb.edu.au; Tel.: +61-3-8344-3017

Received: 30 March 2018; Accepted: 4 May 2018; Published: 7 May 2018

Abstract: There is a significant increase in the number of people with dementia, and the demand for residential support facilities is expected to increase. Providing an appropriate living environment for residents with dementia, which can cater for their specific needs is crucial. Residential aged care design can impact the quality of life and wellbeing of the residents. In this investigation, three recently constructed dementia support facilities in Victoria, Australia are selected for evaluation. Through fieldwork observation, design evaluation and space syntax analysis, the aim of this investigation is to consider the design of these three facilities in the context of current evidence on how the built environment can best accommodate residents with dementia.

Keywords: design for dementia; dementia-friendly; design evaluation; dementia support facilities; residential aged care; built environment; space syntax; wayfinding; behavior cues; orientation cues

1. Introduction

According to the current statistics available from the World Health Organization, around 47 million people have dementia, with 9.9 million new cases being diagnosed every year [1]. Over 425,000 people with dementia are living in Australia, of which about 105,000 people are in Victoria [2,3]. The number of people with dementia in Australia is expected to exceed 1.1 million by 2056. In 2016, over 23% of people with dementia were living in aged care accommodation in Australia [4]. Due to increasing life expectancy, increases in the aging demographic and the significant increase in the number of people with dementia, the demand for residential facilities which provide environments for dementia care is expected to increase. Providing a living environment for residents with dementia, which can cater for their specific needs, is crucial.

Facing the increasing number of people with dementia, the Victorian government developed and published *The Victorian Dementia Action Plan 2014–2018*, which states that designing buildings for people with dementia should be in line with the concept of dementia-friendly environments [5]. A dementia-friendly environment can be defined as "a cohesive system of support that recognizes the experiences of the person with dementia and best provides assistance for the person to remain engaged in everyday life in a meaningful way" [6] (p. 187). The objective is to assist people with dementia to remain socially engaged in everyday life [7].

In this investigation, the following three recently constructed dementia support facilities within residential aged care buildings in Victoria, Australia were selected for field observation: Facility A (inner urban), Facility B (regional), and Facility C (outer urban). All have been managed by the same service provider who has agreed to collaborate and provide access to these facilities. This paper has evolved from a conference presentation by two of the authors [8]. Through field observation, design evaluation, and space syntax analysis, the aim of this investigation is to consider the design of these three facilities in the context of current evidence on how the built environment can best accommodate residents with dementia. In this research, we focused on the design of communal areas rather than bedrooms or staff spaces. The focus was on how shared spaces might be designed to support residents with dementia. The experiences of residents are impacted by many factors such as staffing, treatment, policy, and family, and these were outside the scope of the study.

2. Characteristics of People with Dementia

In order to design physical environments for people with dementia, it is crucial to understand how dementia impacts people. Dementia can change how people perceive their environments. The appropriate physical environments can help compensate for problems associated with dementia [9]. In the case of people with dementia, design decisions should address cognitive impairments, memory loss, confusion, wandering, over/under stimulation, and reduced judgement.

Dementia is a broad term to describe a collection of symptoms that are caused by disorders affecting the brain. Dementia Australia reported that the most common type of dementia is Alzheimer's disease, which affects up to 70% of all people with dementia [10]. According to a *Dementia in Australia* report published by the Australian Institute of Health and Welfare, residents with dementia showed problematic verbal behaviors (such as being verbally disruptive and having paranoid ideation that disturbs others), problematic physical behaviors (including physically threatening or harmful behavior and prolonged physical agitation), severe cognitive skills impairment, wandering behavior, and depression (Table 1) [1].

Table 1. Behavior characteristics of people with dementia (source: AIHW, 2012 [1]).

Behavior Characteristics	Percentage of Residents with Dementia Showing the Behaviors Twice a Day or More
(1) Problematic verbal behaviors	55%
(2) Problematic physical behaviors	50%
(3) Severe cognitive skills	48%
(4) Wandering behavior	27%
(5) Depression	10%

Besides behavioral issues, people with dementia may also encounter increasing difficulties in handling the activities of daily living, including mobility, personal hygiene, toileting, and continence. Currently, there is no cure for dementia, but medications are available to ameliorate symptoms such as agitation and paranoia.

Lubczynski (2014) recommended that when providing care facilities for people with dementia, design decisions should address cognitive impairments, memory loss, confusion, wandering, over/under stimulation, and reduced judgment [11]. Maintaining independence, dignity, a sense of belonging, privacy, and social interaction might also be supported by design [12]. As mentioned by Weisman et al. (1990), "even modest changes in the environments of people of reduced competence may have significant positive consequences" [13].

3. Designing for People with Dementia

3.1. Evidence

The publication titled *Evidence Based Design (EBD) Journal 1: Aged Care* summarised common themes emerging in the process of collating multiple, small, and often non-randomized research projects on design for aging. Evidence for the impact of space, design, and indoor environment quality

(IEQ) on the wellbeing of people with dementia has been growing over recent decades informing a range of guidelines which are largely aligned in their recommendations [14].

In this section, we summarize the key elements of guidelines informing design for people with dementia and the evidence on which they are based. In a later section, we consider how the three dementia support facilities align with the guidelines. The broad, common patterns that have emerged over time and informed the range of guidelines are still evolving. However, there is growing knowledge about what constitutes a home-like environment in terms of dignity, independence, self-expression, scale, familiarity, control, and autonomy. The evidence base has been developed largely by iteratively testing hypotheses through longitudinal post-occupancy evaluation seeking user feedback on design decisions. Sensor-based data collection on IEQ and movement holds promise for future research. More involvement by residents as active participants in design and evaluation teams is needed while understanding the ethical challenges of research involving people with dementia.

People with dementia perceive their environment in ways that differ from people without dementia [15]. The research culture linked to design for dementia has shifted from design for impairment to a more positive focus on designing for remaining abilities, understanding how much of the self is retained even as cognition reduces, and how important this sense of self is to wellbeing [14].

Guidelines developed over the last two decades include the dementia audit tool (DAT), the Dementia Design Checklist developed in Scotland, EVOLVE, Enhancing the Healing Environment (EHE) Assessment Tool, and the Environment Audit Tool (EAT) [16].

3.2. Design Principles

The key elements of the many design guidelines for people with dementia can be summarized as design for [a] homelike settings, [b] orientation, [c] independence, [d] stimulation, [e] safety, and [f] a balance between privacy and community.

[a] Homelike settings are small and familiar, thereby reducing confusion. Homes are an expression of self through personalized furnishings. Personalizing spaces within bedrooms and entries suggests ownership and belonging. Developing a sense of home within a residential aged care setting requires reconciliation of ambiguities in regard to ownership and how private, privileged, and public spaces are defined. Access to smaller semi-private sitting areas where residents can meet with family and guests can help replicate the living space of a home. Smaller clusters of residents also contribute to a sense of home, particularly if dining settings are also domestic in scale or if the dining areas allow for choices similar to choosing a restaurant or cafe. These smaller settings have been linked to increased food intake and social interaction [17,18]. Evidence suggests the residual skills needed for activities of daily living (ADL) are retained for longer when persons with dementia live in a homelike setting [19]. Design can help camouflage those elements that are needed for health care but usually give the appearance of a hospital or an institutional setting. For example, medical files and nursing offices can be back-of-house, thus avoiding the need for nurses' stations.

[b] Orientation using visual clues can reduce the need for mental maps that rely on memory. Clear pathways, memory boards at entry doors to private spaces, landmarks, and destinations help with spatial orientation [20]. A simple network of visually connected spaces helps mobile residents by giving direct lines of sight between bedrooms and destinations. Research indicates that kitchens opening directly onto dining areas facilitates orientation and purpose for residents while enabling care staff to provide unobtrusive oversight. Long corridors with many doors and dead-end corridors should be avoided. New research suggests implicit memory remains intact after other modes of wayfinding are no longer possible, suggesting that unique markers along route and beacon markers at destinations can help orientation [14].

[c] Independence is supported by environments that are familiar and small and where the daily cycles of activity are implicitly understood through visual, aural, and olfactory clues that do not rely on memory and decision-making. Independent functioning is associated with a sense of self whether it is choosing where to be, who to talk with, or what to do, and when to do it. Movement is linked

to independence, as is the choice to be inside and outside. Providing a safe and interesting precinct without the perception of being within a locked environment supports a sense of independence and choice [14]. High contrast settings such as a colored toilet seat or contrasting crockery and table colors can help retain independence and avoid confusion [21].

[d] Stimulus is a complex issue for people with dementia as some may be disturbed by minor environmental stimuli that an unimpaired person would ignore. The challenge is to include positive stimulation that promotes engagement and pleasure while avoiding sensory overstimulation. There is growing evidence that sound has a measurable impact on pleasure [14]. While it is conceivable that light, volume, movement, connection to nature, and smell might also be positive stimuli, there is not yet sufficient research. This issue is linked to orientation and independence. Attention to the indoor environment quality is needed to ensure comfort levels are appropriate for the elderly in terms of air temperature, relative humidity, air movement, light level, ventilation rate, air pollutant concentrations, sound pressure level, room acoustics quality, etc. Spaces for meaningful activities and socialization, as well as withdrawal spaces and a choice of settings, can enable residents to choose their preferred level of stimulation.

[e] Safe and secure spaces can be designed to reduce stressors. A lack of handrails, sharp projections, uneven surfaces and lightweight furniture can increase the risk of injury. Unobtrusive safety reduces the perception of being closed in. Higher lighting levels are necessary as eyes age. Doors in end walls can be attractors for exit-seeking behaviors, whereas side doors can reduce this behavior.

[f] Balancing privacy and community supports wellbeing in a range of ways. Balancing a resident's need for privacy within the context of social connections is difficult within an institutional residential aged care setting. Small spaces near private rooms may provide interstitial or privileged settings shared primarily by a sub-group. Connections into broader familiar communities can be achieved by locating cafes, galleries, or maker spacers at ground level. Other strategies for community connections have been explored elsewhere, particularly in Northern Europe. Childcare has been collocated with aged care, university students have been given accommodation in exchange for a few hours of engagement each week, and in some facilities, pools and services are shared with communities. Each has spatial implications.

4. Observations and Discussion of Findings of the Three Dementia Support Facilities

4.1. General Layout

The three selected dementia support facilities in Victoria, Australia were built in the 2010s (Facility A in 2014, Facility B in 2015 and Facility C in 2017). All of them provide single bedrooms with ensuites and small sitting areas. Among them, Facility B has the capacity to accommodate up to 34 residents. The 34 bedrooms are grouped into four wings, with eight bedrooms in two wings, and seven and nine bedrooms in the remaining two wings. Bedrooms are located on both sides of the corridors, with a maximum length of five bedrooms. Communal spaces at the central portion link the four corridors together. On the northeast side of the facility, there is an outdoor garden (Figure 1).

Facility A has the smallest capacity among the three facilities catering for 13 residents, with seven bedrooms on one side (House 1) and six bedrooms on the other side (House 2). Bedrooms are in L-shaped configuration in House 1 and in linear arrangement in House two with a corridor of three bedrooms in length. House 1 and House 2 are separated by an activity room, but are open to the same covered terrace outside. (Figure 2, left). Facility C has a slightly larger capacity than Facility A and can cater for 17 residents. It has eight bedrooms on one side (House 1) and nine bedrooms on the other side (House 2). Corridors in each house are in a T-shape configuration. House 1 and House 2 are connected by a service corridor for staff access. Each House opens to an outdoor terraced garden. (Figure 2, right).

All these three dementia support facilities have homelike settings with small sitting areas with views to facilitate social interaction and provide unrestricted access to safe exteriors, either secured gardens or balconies. Facility B is located within a retirement village in a regional area, whereas Facilities A and C are located in residential aged care buildings with cafes on ground floor, so Facilities A and C have stronger connections to local communities than Facility B.

A comparison table of the general layout of these three facilities is shown in Table 2. For any variable, there are the three following levels: *, **, and ***, which is indicative of relative assessment, according to the level of engagement of a particular variable, ranging from good (*), better (**) to the best (***).

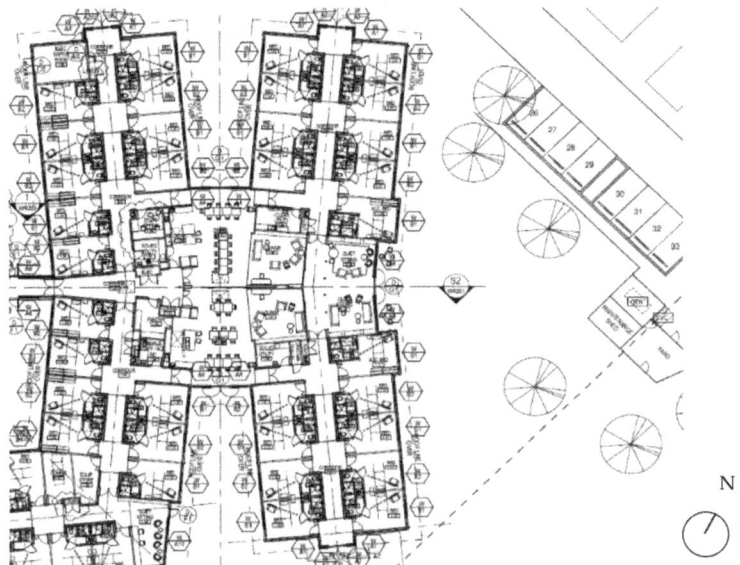

Figure 1. Floor plan of Facility B.

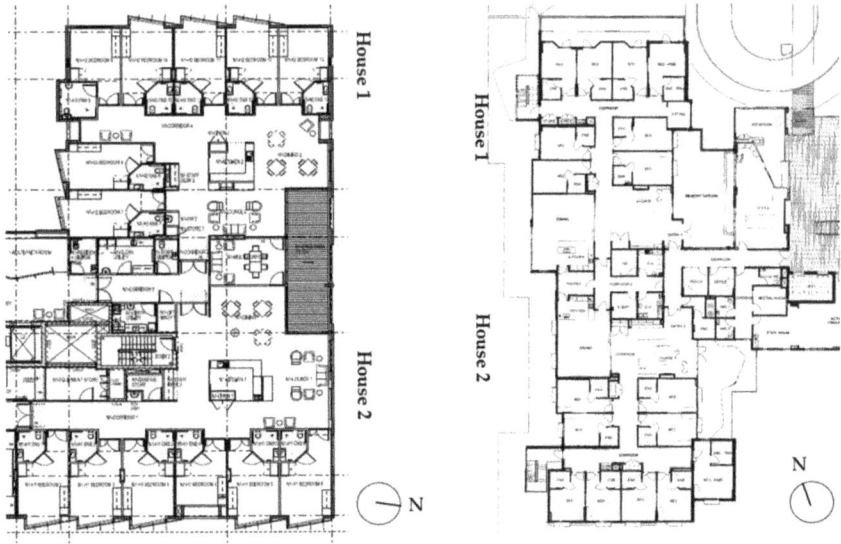

Figure 2. Floor plans of Facility A (**left**) and Facility C (**right**).

Table 2. Comparison of the general layout.

Design Principles	Facility A	Facility B	Facility C
Homelike: Small	***	*	**
Homelike: Access to small sitting areas	***	***	***
Orientation: Simple layout	***	**	***
Orientation: Short corridors	***	*	**
Balancing privacy and community: Connection to community	***	**	***

Notes: * = Good; ** = Better; *** = Best.

4.2. Dining Area and Kitchen with Domestic Setting

The dining hall at Facility B can be divided into two halves by sliding partitions, resulting in two smaller dining areas to cater for 17 residents on one side. This offers adaptive spatial usage to cope with the needs of the residents. Since the gathering of 34 people together at the same place may create too high a noise level, resulting in overstimulation, agitation, and confusion to some residents, the flexibility of spatial subdivision can reduce the possibility of disruptive behaviors during mealtimes [22]. The dining hall offers visual and physical access to gardens on both sides providing spatial orientation cues and helpful stimulation to residents. Windows at both the southeastern side and northwestern side can also allow natural light to enter to the interior with control mechanisms against glare.

At Facility B, the domestic setting of the kitchen is the focal point of the dining area. It does not replace the full-service kitchen, but breakfast preparation, beverage making, and dessert baking can contribute to the domestic ambience of the space, reducing the image of the overall institutional setting. The distinctive smell of food during meal preparation provides olfactory cues to residents. The kitchen next to the dining area also facilitates the staff to cater for personal dietary requirements and allows residents to make choices, especially during breakfasts, so that they may feel more in control of their lives, which has positive implications for the sense of competence and self-esteem of people with dementia. The kitchen is not merely a food preparation area, but also "a practical and non-institutional alternative to the traditional nurses' station" [10]. Staff at the kitchen enjoy an unobstructed view of the dining area, adjacent living areas and the outdoor garden beyond, which offers informal surveillance and ease of monitoring of the residents.

A similar domestic kitchen arrangement is also provided at Facility C. Compared to the open plan kitchens at Facilities B and C, the domestic kitchen at Facility A is more enclosed with glass doors to prevent unauthorized entry (Figure 3, Table 3).

Figure 3. Kitchen with domestic setting: Facility A (**left**), Facility B (**middle**), and Facility C (**right**).

Table 3. Comparison of dining areas and kitchens with the domestic setting.

Design Principles	Facility A	Facility B	Facility C
Homelike: Access to small dining settings	***	**	***
Stimulus: Olfactory cues during meal preparation	***	***	***
Independence: Choice of spaces with views	***	***	***

Notes: * = Good; ** = Better; *** = Best.

4.3. Outdoor Gardens

Among the three dementia support facilities, the outdoor garden at Facility B is the biggest. Doors opening to the garden are unlocked during the daytime, which enables residents going outside as one of their choices. This may lead to the decrease in negative aggressive behaviors of the residents [23]. In fact, a well-designed garden is a therapeutic environment for people with dementia as it can provide visual, tactile, olfactory, and auditory stimulation through the combination of natural landscape, fragrance, sunlight, wind, and birds. The timber trellis at the entrance of the garden serves as an iconic structure for residents' spatial orientation. If more interest points can be provided along the looped path and appropriate shelters can be erected to protect seating areas from excessive solar radiation exposure and strong wind, this may attract more residents to visit the garden. Wheelchair-accessible raised planting beds can also be provided to allow residents with remaining abilities to participate in gardening.

The open terraced garden at Facility C is relatively small and there was not much planting at the time of visit prior to occupation by residents. The garden has potential to be a source of sensory stimulation to residents if it is properly landscaped. Gardens at Facilities B and C face northeast and southeast respectively. Both of them can capture favorable morning sunlight, encouraging residents to go outside. The outdoor activity area at Facility A is the smallest with a covered terrace and limited planting. It faces north, but due to its openings on one side and its close proximity to the adjacent building, solar radiation exposure is unavoidably affected. Solar penetration to the communal space behind the covered terrace is further reduced due to the setback from the façade (Figure 4, Table 4).

Figure 4. Outdoor gardens: Facility A (**left**), Facility B (**middle**), and Facility C (**right**).

Table 4. Comparison of outdoor gardens.

Design Principles	Facility A	Facility B	Facility C
Independence: Unrestricted access to safe exteriors	***	***	***
Stimulus: Landscape, fragrance and sunlight	*	***	**

Notes: * = Good; ** = Better; *** = Best.

4.4. Corridors

In Facility A, there are memory boxes outside residents' rooms along the corridors. The inclusion of personal objects in the memory boxes, such as photos and other artefacts, facilitates residents with dementia to reinforce their long-term memory and reflect upon their past experiences within their remaining capabilities. This can personalize the institutional setting and enhance the sense of identity by creating a familiar environment and serve as an effective orientation cue for wayfinding [24]. Displaying personal objects along the corridor may also stimulate social interaction and conversation among residents and enable the staff to have better understanding of the residents about their stories and preferences [25]. However, the corridors at Facilities B and C only have pictures hanging on walls and color contrast without memory boxes (Figure 5, Table 5).

Table 5. Comparison of corridors.

Design Principles	Facility A	Facility B	Facility C
Orientation: Visual cues	***	**	**
Safety: Well-positioned handrails	***	***	***

Notes: * = Good; ** = Better; *** = Best.

Figure 5. Corridors: Facility A (**left**), Facility B (**middle**), and Facility C (**right**).

5. Space Syntax Analysis of the Three Dementia Support Facilities

Apart from design evaluation through field observation, space syntax analysis was applied for comparing the configurations of the three facilities. Space syntax relies on the use of mathematics of graph theory to measure the spatial and social properties of plans for tracing the underlying layer of space that accommodates real conditions of human movement, access, and surveillance [26–28]. Spatial variables such as visibility (visual connectivity, openness, visual cues) and the relative depth of spaces (proximity, accessibility) can influence the social interactions, spatial orientation, and wayfinding abilities of people with dementia [29–31]. In this investigation, the depthmapX software developed by the Space Syntax Laboratory at the University College London (UCL) was employed to accomplish visibility graph analysis, isovist analysis, and step depth analysis for comparison and discussion [32].

5.1. Visibility Graph Analysis

Visibility graph analysis is a common computational approach of space syntax based on two-dimensional representations of space. The properties of the plans are abstracted and mathematically analyzed to reveal the connectivity of different spaces [33]. Full-height partitions and walls are taken as boundaries, while doors and openings are considered as connection points. Visibility graphs are colored, ranging from red to dark blue to represent different degrees of connectivity.

As shown in Figure 6, the four wings of the Facility B have low connectivity values (dark blue), which are more visually and socially isolated. On the contrary, both Facilities A and C have higher connectivity, especially their communal spaces (living and dining areas), which can facilitate social interaction among users with the ease of physical and visual access (Table 6). The least connected spaces are bathrooms and service rooms, as represented by dark blue on the analysis diagrams.

Table 6. Comparison of connectivity of spaces.

Design Principles	Facility A	Facility B	Facility C
Orientation: Visually connected spaces	***	**	*

Notes: * = Good; ** = Better; *** = Best.

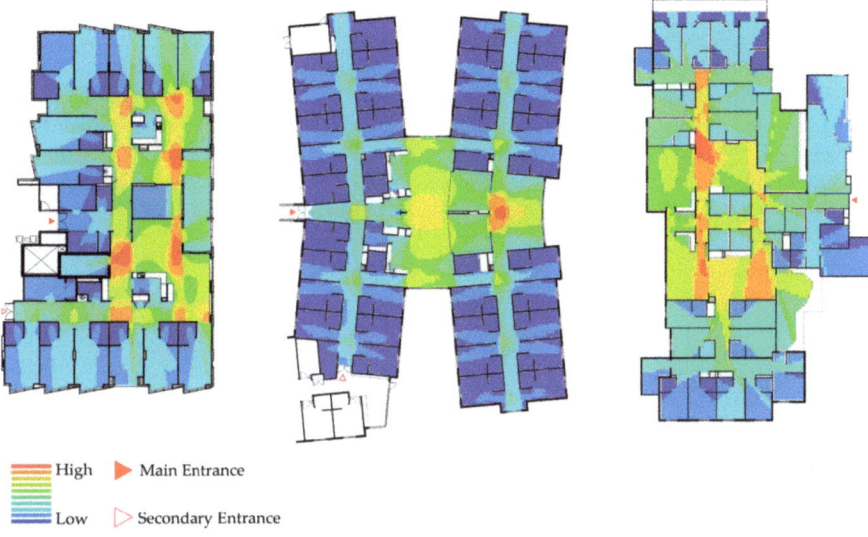

Figure 6. Visibility graph analysis: Facility A (**left**), Facility B (**middle**), and Facility C (**right**).

5.2. Isovist Analysis

Isovist analysis was initially developed by Tandy in 1967 for landscape surveys [33]. By defining 'isovist' as a 'set of all points visible from a given vantage point in space', Benedikt introduced an analytic method for quantitative descriptions of spatial environments in 1979 [34,35]. This is an effective tool to illustrate the visibility of a particular point in the layout plan. Since the domestic kitchen, found at each dementia support facility, is a key focal point in the communal space, it is used for developing the isovist analysis diagrams in Figure 7.

Figure 7. Isovist analysis: Facility A (**left**), Facility B (**middle**) & Facility C (**right**).

The domestic kitchen at Facility A is strategically located at the center of the whole layout, which provides the carers the ease of surveillance for ensuring the safety and wellbeing of residents with dementia. Visibility may be further enhanced if the domestic kitchen is not enclosed by full-height partitions.

The visibility of domestic kitchens at both Facilities B and C is restricted to communal spaces and cannot reach the corridors. On the other hand, the layout configuration of Facility B enables the carers at the domestic kitchen to be visually connected to different types of communal spaces (dining areas, living areas, and lounges facing the garden outside). Comparatively, the visibility of the domestic kitchen at Facility C is more confined due to its location at the corner of the layout (Table 7).

Table 7. Comparison of direct lines of sight.

Design Principles	Facility A	Facility B	Facility C
Orientation: Direct lines of sight	***	**	*

Notes: * = Good; ** = Better; *** = Best.

5.3. Step Depth Analysis

In view of the wandering behavior and cognitive impairment of residents with dementia, it is preferable for dementia support facilities to have lower relative depth for ease of wayfinding and spatial orientation. Step Depth Analysis is an effective visual tool to illustrate the relative depth of the spaces. Different types of spaces on the layout plan are firstly labelled. Spaces with different levels of step depth are represented in the layout by different colours, ranging from red, orange, green, cyan, to purple (Figure 8). The physical connections of different spaces on the layouts are then represented by a tree diagram using the main entrance as Level 0 (Figure 9). In general, communal spaces close to the main entrance have lower step depth, whereas bedrooms have higher step depth.

Among the three dementia support facilities, Facility A has the least step depth (bedrooms have only Level 2 step depth), while Facility C has the greatest step depth (all bedrooms have Level 3 step depth). Due to the balcony outside some bedrooms, the step depth of Facility C can even reach Level 4 (Table 8).

Figure 8. Relative depth analysis layout: Facility A (**left**), Facility B (**middle**), and Facility C (**right**).

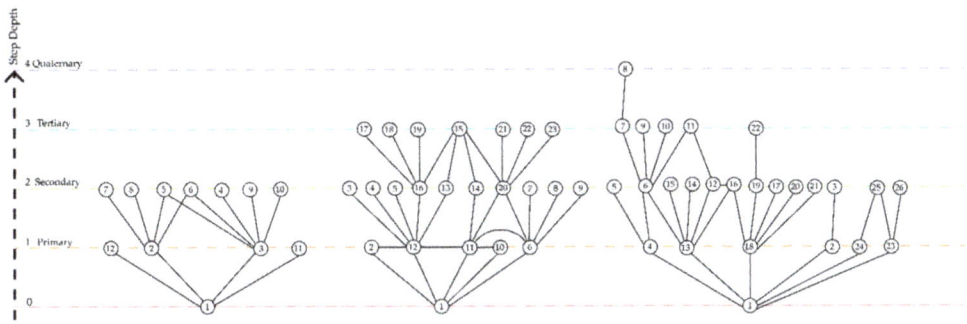

Figure 9. Tree diagrams: Facility A (**left**), Facility B (**middle**), and Facility C (**right**).

Table 8. Comparison of relative depth of spaces for wayfinding.

Design Principles	Facility A	Facility B	Facility C
Orientation: Depth of spaces for wayfinding	***	**	*

Notes: * = Step depth Level 4; ** = Step depth Level 3; *** = Step depth Level 2.

6. Conclusions

What are the design lessons learnt from this investigation? Through fieldwork observation, design evaluation, and space syntax analysis, the general building layouts of the three selected dementia support facilities were compared. How each design addressed current guides on design for dementia was explored in terms of providing a homelike setting and designing for better spatial orientation, independence, stimulus, and safety, as well as balancing privacy with community. Key design factors were identified and appropriate provisions within the facilities were discussed, including the following:

(1) visual access and clear sight line within the domestic ambience of the space
(2) use of kitchen with domestic setting as an alternative to the traditional nurses' station
(3) adaptive spatial usage to cope with disruptive behaviors of residents with dementia
(4) outdoor activity spaces for connection to nature, particularly in Facilities B and C
(5) overall layout with higher visual connectivity for enhancing social interaction and lower step depth for ease of spatial orientation

Design evaluation of these three dementia support facilities can be summarized in Table 9:

Table 9. Design evaluation of the three dementia support facilities.

Design Principles	Facility A	Facility B	Facility C
[A] Homelike			
Small	***	*	**
Access to small sitting areas	***	***	***
Access to small dining settings	***	**	***
[B] Orientation			
Simple layout	***	**	***
Short corridors	***	*	**
Visual cues	***	**	**
Visually connected spaces	***	**	***
Direct lines of sight	***	**	*
Depth of spaces for wayfinding	***	**	*

Table 9. *Cont.*

Design Principles	Facility A	Facility B	Facility C
[C] Independence			
Choice of spaces with views	***	***	***
Unrestricted access to safe exteriors	***	***	***
[D] Stimulus			
Olfactory cues during meal preparation	***	***	***
Landscape, fragrance, and sunlight	*	***	**
[E] Safety			
Well-positioned handrails	***	***	***
[F] Balancing privacy and community			
Connection to community	***	**	***

Notes: * = Good; ** = Better; *** = Best.

The research on the impact of design of living environment on the quality of life and wellbeing of residents with dementia is ongoing. Further ethnographic analysis including photo elicitation and semi-structured interviews with carers and relevant design practitioners will be carried out to collect and collate their feedback. This can inform the design strategies of future dementia support facilities to suit the specific needs of people with dementia.

Author Contributions: Conceptualisation, methodology and writing-original manuscript, H.-w.C.; Section 3, C.N.; Data collection and analysis, C.M.M.W.; Section 1, N.M.; Space syntax diagrams and analysis, J.W.; Final review and editing, L.A.

Funding: This research was funded by an Early Career Researcher Grant provided by the Faculty of Architecture, Building and Planning, the University of Melbourne.

Acknowledgments: The authors wish to thank Australian Unity for providing access to the three dementia support facilities in Victoria, Australia for this initial investigation and allowing relevant floor plans and images to be published.

Conflicts of Interest: The authors declare no conflict of interest.

References

1. AIHW. Dementia in Australia, Australian Institute of Health and Welfare, Canberra, 2012. Available online: http://www.aihw.gov.au/WorkArea/DownloadAsset.aspx?id=10737422943 (accessed on 6 August 2017).
2. Dementia Australia. Dementia: Key Facts and Statistics 2018, 2018. Available online: https://www.dementia.org.au/files/documents/Key-facts-and-statistics.pdf (accessed on 30 March 2018).
3. Dementia Australia. Dementia Statistics for Victoria, 2017. Available online: https://www.dementia.org.au/statistics/vic (accessed on 30 March 2018).
4. Brown, L.; Hansnata, E.; La, H.A. *Economic Cost of Dementia in Australia 2016–2056*; University of Canberra: Canberra, Australia, 2017.
5. Department of Health. *Supporting People with Dementia and their Families and Carers: Victorian Dementia Action Plan 2014–18*; State Government of Victoria: Melbourne, Australia, 2014.
6. Davis, S.; Byers, S.; Nay, R.; Koch, S. Guiding design of dementia friendly environments in residential care settings: Considering the living experiences. *Dementia* **2009**, *8*, 185–203. [CrossRef]
7. Dementia Training Australia. Module 8: Creating Dementia Friendly Environments, 2017. Available online: https://www.dta.com.au/wp-content/uploads/2017/02/Module-8-Creating-dementia-friendly-environments.pdf (accessed on 30 March 2018).
8. Chau, H.W.; Newton, C. A Pilot Study of Design Evaluation of Three Memory Support Residential Facilities in Victoria. In *Back to the Future: The Next 50 Years*, Proceedings of the 51st International Conference of Architectural Science Association (ANZAScA), Wellington, New Zealand, 29 November–2 December 2017; Schinabel, M.A., Ed.; Victoria University of Wellington: Wellington, New Zealand, 2017; pp. 763–772.

9. Fleming, R. Towards providing Better Care Planning and Environmental Design for People with Dementia in Residential Aged Care. Ph.D. Thesis, School of Nursing, Midwifery and Indigenous Health, University of Wollongong, Wollongong, Australia, 2013. Available online: http://ro.uow.edu.au/theses/3926 (accessed on 30 March 2018).

10. Dementia Australia. Alzheimer's Disease, 2017. Available online: https://www.dementia.org.au/about-dementia/types-of-dementia/alzheimers-disease (accessed on 22 April 2018).

11. Lubczynski, S. Architecture as Third Skin Spatial Dimensions of Stimuli for Dementia Care. Master's Thesis, Ryerson University, Toronto, ON, Canada, 2014.

12. Prokopová, A. Living Knowhere: Research and Design on Dementia and Architecture. Master's Thesis, Faculty of Architecture and The Built Environment, TU Delft, Delft, The Netherlands, 2015.

13. Weisman, G.; Cohen, U.; Day, K.; Meyer, G. *Programming and Design for Dementia: Development of a 50 Person Residential Environment*; University of Wisconsin-Milwaukee: Milwaukee, WI, USA, 1990.

14. O'Brien, D. (Ed.) *Evidence Based Design (EBD) Journal 1: Aged Care*; 2014; Volume 1, pp. 1–69. Available online: http://ebdjournal.com/journals/aged-care (accessed on 1 August 2015).

15. Fay, R.; Fleming, R.; Robinson, A. Design for Dementia: Sustainability and Human Wellbeing. In *On the Edge, Proceedings of the 44th Annual Conference of the Australian and New Zealand Architectural Science Association, Auckland, New Zealand, 24–26 November 2010*; Unitec Institute of Technology: Aukland, New Zealand, 2010.

16. O'Malley, M.; Innes, A.; Wiener, J.M. Decreasing spatial disorientation in care-home settings: How psychology can guide the development of dementia friendly design guidelines. *Dementia* **2017**, *16*, 315–328. [CrossRef] [PubMed]

17. Calkins, M.P. Evidence-based long term care design. *NeuroRehabilitation* **2009**, *25*, 145–154. [PubMed]

18. Reed, P.S.; Zimmerman, S.; Sloane, P.D.; Williams, C.S.; Boustani, M. Characteristics associated with low food and fluid intake in long-term care residents with dementia. *Gerontologist* **2005**, *45*, 74–80. [CrossRef] [PubMed]

19. Fleming, R.; Crookes, P.A.; Sum, S. *A Review of the Empirical Literature on the Design of Physical Environments for People with Dementia*; The University of Wollongong: Wollongong, Australia, 2008. Available online: http://ro.uow.edu.au/hbspapers/2874/ (accessed on 30 March 2018).

20. Milke, D.L.; Leask, J.; George, C.; Ziolkowski, S. Eight Years of Data on Residents in Small Dementia-Care Settings Suggest Functional Performance Is Maintained. *J. Hous. Elder.* **2015**, *29*, 298–328. [CrossRef]

21. Van Hoof, J.; Kort, H.S.; Van Waarde, H.; Blom, M.M. Environmental interventions and the design of homes for older adults with dementia: An overview. *Am. J. Alzheimers Dis. Other Dement.* **2010**, *25*, 202–232. [CrossRef] [PubMed]

22. Schwarz, B.; Chaudhury, H.; Tofle, R.B. Effect of Design Interventions on a Dementia Care Setting. *Am. J. Alzheimers Dis. Other Dement.* **2004**, *19*, 172–176. [CrossRef] [PubMed]

23. Namazi, K.H.; Johnson, B.D. Pertinent Autonomy for Residents with Dementias: Modification of the Physical Environment to enhance Independence. *Am. J. Alzheimers Dis. Other Dement.* **1992**, *7*, 16–21. [CrossRef]

24. Namazi, K.H.; Rosner, T.T.; Rechlin, L. Long-term memory cuing to reduce visuo-spatial disorientation in Alzheimer's Disease patients in a special care unit. *Am. J. Alzheimers Dis. Other Dement.* **1991**, *6*, 10–15. [CrossRef]

25. Kovach, C.; Weisman, G.; Chaudhury, H.; Calkins, M. Impacts of a therapeutic environment for dementia care. *Am. J. Alzheimers Dis. Other Dement.* **1997**, *12*, 99–110. [CrossRef]

26. Hiller, B. *Space is the Machine: A Configurational Theory of Architecture*; Cambridge University Press: Cambridge, UK, 1999.

27. Hiller, B.; Hanson, J. *The Social Logic of Space*; Cambridge University Press: Cambridge, UK, 1984.

28. Lee, J.H.; Ostwald, M.J.; Lee, H. Measuring the spatial and social characteristics of the architectural plans of aged care facilities. *Front. Architect. Res.* **2017**, *6*, 431–441. [CrossRef]

29. Day, K.; Carreon, D.; Stump, C. The therapeutic design of environments for people with dementia: A review of the empirical research. *Gerontologist* **2000**, *40*, 397–416. [CrossRef] [PubMed]

30. Ferdous, F.; Moore, K. Field observations into the environmental soul: Spatial configuration and social life for people experiencing dementia. *Am. J. Alzheimers Dis. Other Dement.* **1995**, *30*, 209–218. [CrossRef] [PubMed]

31. Faith, V. Designing for Dementia: An Assessment of the Impact of the Physical Environment on Wayfinding Success for Residents in Long Term Care Settings. Ph.D. Thesis, Queen's University Belfast, Belfast, UK, 2014.

32. Turner, A. Depthmap: A Program to perform Visibility Graph Analysis. In Proceedings of the Third International Space Syntax Symposium, Atlanta, GA, USA, 7–11 May 2001; Peponis, J., Wineman, J., Bafna, S., Eds.; Georgina Institute of Technology: Atlanta, GA, USA, 2001; pp. 31.3–31.9.
33. Tandy, C.R.V. The Isovist Method of Landscape Survey. In *Symposium: Methods of Landscape Analysis*; Murray, H.C., Ed.; Landscape Research Group: London, UK, 1967; pp. 9–10.
34. Benedikt, M.L. To take hold of space: Isovists and isovist fields. *Environ. Plan. B* **1979**, *6*, 47–65. [CrossRef]
35. Turner, A.; Doxa, M.; O'Sullivan, D.; Penn, A. From isovists to visibility graphs: A methodology for the analysis of architectural space. *Environ. Plan. B* **2001**, *28*, 103–121. [CrossRef]

 © 2018 by the authors. Licensee MDPI, Basel, Switzerland. This article is an open access article distributed under the terms and conditions of the Creative Commons Attribution (CC BY) license (http://creativecommons.org/licenses/by/4.0/).

Article

Patients' Perspectives on the Design of Hospital Outpatient Areas

Yisong Zhao [1] **and Monjur Mourshed** [2,*]

[1] School of Architecture, Building and Civil Engineering, Loughborough University, Loughborough LE11 3TU, UK; easonbeingcrazy@gmail.com
[2] School of Engineering, Cardiff University, Cardiff CF24 3AA, UK
* Correspondence: MourshedM@Cardiff.ac.uk; Tel.: +44-29-2087-4847

Received: 23 October 2017; Accepted: 1 December 2017; Published: 5 December 2017

Abstract: There is a growing interest among healthcare managers and designers in moving towards a 'patient-centred' design of health and care facilities by integrating patient perceptions and expectations of the physical environment where care takes place. Increased interests in physical environments can mostly be attributed to our improved understanding of their role in patients' health outcomes and staff productivity. There is a gap in the literature on users' perspectives on physical settings in the context of healthcare. Moreover, the connection of care services with the design of the facility is often overlooked partly due to the lack of evidence. This research was aimed at filling the gap by exploring outpatients' perspectives on design factors related to the areas frequented by them, e.g., hospital waiting areas. A 16-item questionnaire was conducted among randomly selected outpatients in two hospitals in Qingdao, China, with a response rate of 84.3%. Five principal factors were identified: sensory; lighting and thermal; facilities; spatial; and seating design, which agreed with the literature. Non-parametric tests were applied to assess variances in constructed principal dimensions concerning demographic variables. Female outpatients were found to be more perceptive of the 'sensory design' factors than males. The number of previous visits to the hospital was found to be associated with 'spatial' and 'seating design' factors, while respondents' age had an association with 'sensory' and 'seating design' factors. Respondents ranked 'noise' and 'air freshness' and 'cleanliness' as highly important.

Keywords: healthcare design; outpatients' perspectives; waiting areas; patient-centred design (PCD); patient-centred care (PCC)

1. Introduction

Patient healing is a complex and dynamic process, during which the role of the physical environment has been recognised and emphasised by many researchers due to its influence on patient health outcomes and wellbeing [1–3]. The interconnections between the characteristics of the physical environment and patient health outcomes emphasise the importance of the physical environment design in creating a healing environment [3–5]. Moreover, physical environment factors during both design and operation of buildings are important for sustainability [6], as well as for meeting the needs and expectations of stakeholders. There is, therefore, a growing interest in the patient-centred architectural design of healthcare facilities among researchers and service providers [7,8]. Patient-centred design (PCD) is a process involving design and evaluation that pays attention to facility users [9,10]; that is, human factors that affect the outcome of the healing process. It requires healthcare architects/designers making an effort to shape and reshape the healing environment, addressing patients' needs to provide satisfying healing experience and achieve desired outcomes of perceived service quality [11]. Traditionally, the quality of healthcare is evaluated by professional practice standards, but, over the last decade, measurement of patient satisfaction has become popular [12]. With the aim of obtaining patients' perspectives about their care, it is increasingly being accepted as an essential indicator of the quality of care [13,14]. However, no consensus exists on which dimensions of care should be evaluated to measure patient satisfaction [15], primarily due to the multi-dimensionality of the concept of patient satisfaction. It has been observed that general patient satisfaction and patient perspectives of healthcare quality are sometimes interpreted interchangeably, but at closer inspection, they are different [16,17]. Patient satisfaction is often described as the patient's subjective experience during their provision of healthcare. It reflects the extent that their expectations and obligations of service standards are met [18]. Usually, when patients perceive that one or more of their expectations for care have been unmet, satisfaction as a whole suffers. Satisfaction reflects much more of personal preferences which are viewed as a broader concept while patient perspectives of service quality focus on dimensions of service. Although these are two different concepts, patients' perspectives of service quality and satisfaction have certain things in common [19].

Various methods to improve healthcare quality have been explored in the past. Researchers investigated patients' perspectives on diverse aspects of care service: waiting time [20,21], interaction and interpersonal skills [21,22], professionalism [23,24], occupancy [25], patient preferences and expectations [26,27], coordination of care [22,28], education and information provision [22,29,30], emotional support [31,32], and quality of medical care [33,34].

Currently, there is little research into patients' perspectives associated with built environment design factors in healthcare. Few researchers have explored the nature and the range of factors that patients consider important to their health and wellbeing. However, the perception of these factors on the design of healthcare facilities and how these can be better integrated into the process of facility design have largely remained overlooked [5]. In addition, as patients have become better educated about healthcare, their perspectives and expectations are changing as well, some previous aspects of measured attitudes may not adequately interpret patient's changing needs [17]. Therefore, this research is aimed to assess outpatients' perspectives of the physical waiting environment, investigating their opinion of a range of important hospital design indicators and reflecting on the building design process.

The rest of the paper is organised as follows. The methods applied in this paper for the development of the instrument and the conduct of the survey are discussed. Descriptive and statistical analyses of the obtained data are discussed next, followed by a contextual discussion. The article ends with a summary of findings and concluding remarks.

2. Methodology

2.1. Overview of Questionnaire Development

The questionnaire development followed four phases. First, the items of the questionnaire were generated based on an extensive review of literature and industry guidelines, conducted from January to May 2009. The purpose of the review was to determine the following:

- Factors related to the design of the physical environment in healthcare facilities;
- Outpatients' perspectives of the physical environment; and
- The physical-environment factors that affected outpatients' outcomes.

Keyword searches were conducted on the following databases: PubMed, ScienceDirect, Web of Science, Scopus, Ovid MEDLINE, the Cochrane Library and Design and Applied Arts Index. This enabled the first-step filtering of literature, which was refined further with keyword searches that were related to the scope and methods; for example, outpatient questionnaire, survey, physical environment, perspectives, healthcare waiting areas. Non-electronic sources were also consulted to identify potential sources for inclusion in the review. The filtered sources, both electronic and non-electronic, were first categorised based on their adopted methods and findings. Relevant design indicators were identified from this systematic review of the literature.

Second, one of the authors visited the two participating hospitals four times and carried out interviews with ten outpatients. A focus group (one-hour session) involving two outpatients, two care providers (nursing staff) and one administrative staff was conducted. In both the focus group and interviews, the objective was to compare the findings of the literature review with participants' perspectives of the waiting environment.

Third, a draft questionnaire was developed by incorporating the findings from the first and second stages. The questionnaire was first produced in English and then translated into Chinese for respondents' convenience. The draft questionnaire was then evaluated in a pilot study to analyse the comprehensibility and clarity of the items and attributes related to the psychometric properties of the instrument. The participating outpatients ($n = 19$) of the pilot study were asked to state any deficiencies of the content of the questionnaire, other potential sources of perspectives and significance of each item. The pilot study resulted in an amended final questionnaire with improved content validity.

The final structure of the questionnaire included 16 questions to rate the perspectives of the importance with regards to the dimensions of hospitals' waiting environment. Respondents were asked to rate their perspectives of an item on a Likert-type response scale, ranging from least important, unimportant, neither important nor unimportant, important and most important, transformed into a scale between 1 and 5—a higher score indicating a higher level of importance for the item. Demographic information such as age and gender were obtained from the participants. Data regarding number of visits, type of the appointment and the visited hospital department were recorded as well.

2.2. Ethical Approval and Study Sample

The ethical approval for the study was obtained in two stages. First, an ethical approval was obtained from the UK academic institution where the authors were based. Second, the research committees of the two participating hospitals gave approval to the study. Written consent was obtained for each interview carried out. The anonymity of respondents has been preserved, except when explicit permission was given to use titles or names. The study was conducted among outpatients in two Chinese hospitals in Qingdao, a coastal city in East China. The hospitals were chosen for this research because they serve a relatively large number of staff and patients offering us an opportunity to select the study sample from a wider background, and for the on-site world standard facilities so that the study findings can be interpreted against other international studies. One of the hospitals is affiliated with a medical college, and the other is the largest general hospital in the city. These two hospitals employ a total of approximately 5900 staff and have around 4000 beds. Respondents were selected to participate in the survey by random sampling from different outpatient departments. All participated respondents were over 18 years old, and they were informed in writing through an introduction to the survey section that the survey was voluntary, and the confidentiality of the data would be retained.

2.3. Data Collection

Some particular holidays (e.g., National day and Spring Festival) in China may create potential bias in the use of healthcare facilities due to festival decorations and lighting, and bring bias in outpatients' perspectives to the physical environment. Data for this study were, therefore, collected between 12 and 26 August 2009, a period in which there were no special holidays in China. The surveyed outpatients were randomly selected from each floor in the outpatient department in both hospitals, from 8:00 a.m. to 5:00 p.m., Monday to Friday, during the two-week study period, to capture all time stages of outpatients' visits. The researcher distributed the questionnaires to the sampled outpatients and explained the purpose of the survey. Informed consent was obtained from each participant in the study. All the survey items were completed by either the outpatients or their guardians (guardians were used if the sampled patients had difficulties in writing). The researchers also verified the questionnaires for completeness and correctness for completion. These completed questionnaires were collected on the spot when finished. A total 337 outpatients from the two Chinese hospitals completed the questionnaires effectively out of 400 distributed, and the results were included in the study. The response rate was 84.3%.

2.4. Statistical Analysis

Most statistical analyses have been performed with IBM SPSS Statistics version 22.0 for Windows. Descriptive statistics on the item and scale frequencies, percentages, means and standard deviations (SD) were computed. Demographic and other related data were also analysed descriptively by computing frequencies and percentages. Internal consistency reliability was assessed via Cronbach's coefficient alpha [35], with $\alpha \geq 0.70$ as the recommended value since this study involved the comparison of groups of respondents [36]. The coefficient $\alpha \geq 0.70$ was regarded as acceptable, $0.80 \geq \alpha > 0.70$ as respectable and $\alpha \geq 0.80$ as very good.

Previous research suggested that a questionnaire with multi-item scales can be used to reduce random sources of errors to represent the theoretical concept [36]. This study, therefore, employs Principal Components Analysis (PCA) to identify the underlying structure characterising a set of highly correlated variables. Varimax rotation was applied to the principal component analysis (PCA) results, guiding the number of factors to be extracted. Items were included in the factors if there were substantial loadings (≥ 0.40). In the case of multiple loadings of an item on different factors, it was included in the factor with which the item had more conceptual relationship. The factors from the PCA results were easier to label and had good correspondence with other studies. After this, good construct validity and internal consistency were established for the questionnaire. Bartlett's test of sphericity was used to identify significant correlation between items. The Kaiser–Meyer–Olkin procedure for measuring sample adequacy was applied.

Chi-square and non-parametric tests were applied to analyse demographic effects and relationships among constructed dimensions. Statistically significant differences in perspectives between genders and appointment types were tested via Mann–Whitney U-test. Differences between the age groups (18–25, 26–35, 36–50, and >50 years) and visit times (1–2, 3–4, 5–10, and >10 times) were analysed using Kruskal–Wallis test with a $p < 0.05$ taken as statistically significant. Mann–Whitney U-test with a reduced p-value ($p < 0.01$) was used as a post hoc test to avoid the risk of finding significant differences by chance [37].

3. Results and Analysis

3.1. Respondents' Characteristics

Demographic and other clinical information from the respondents is given in Table 1. Among 337 surveyed outpatients, 124 (36.8%) were male and 213 (63.2%) were female. More than half of the male respondents were aged between 26 and 35, nearly a quarter of male respondents were aged between 36 and 50, 16 respondents were aged between 18 and 25 years and 14 male respondents were >50. Similarly, most female respondents were aged between 26 and 35 years with only 15 female participants >50. Male respondents visited the hospital less frequently than female. Table 1 shows 77.5% of female respondents have visited the hospital more than twice compared with a smaller number of 66.1% of male respondents. Most of the respondents pre-arranged their visits while only seven male respondents were admitted as an emergency. Outpatients were selected from 22 departments across the hospitals; the department of general surgery ($n = 79$) and respiratory ($n = 59$) represent the relatively higher number of returned questionnaires than other departments. The diversity of different departments ensured a wide range of respondents were represented in the study.

A descriptive analysis of the design indicators is given in Table 2, which shows the percentage of responses at each choice of the five-point scale. Mean and standard deviations (*SD*) of responses are computed for each design indicator. The questionnaire items are sorted in descending order, based on the mean response score. Standard deviations are generally small for higher mean response scores (e.g., cleanliness; mean = 4.55, SD = 0.565) and relatively greater for lower mean scores (e.g., presence of coordinated art objects; mean = 3.18, SD = 0.943).

Table 1. Demographic information of the respondents.

Variable	Scale	Male (124)	Female (213)	*p* Value [†]	Total (%)
Age (year)				0.019	
	18–25	16	51		19.9
	26–35	64	84		43.9
	36–50	30	63		27.6
	>50	14	15		8.6
Number of visits				0.136	
	1–2	42	48		26.7
	3–4	37	71		32.0
	5–10	21	39		17.8
	>10	24	55		23.4
Appointment type				<0.001	
	Emergency	7	0		2.1
	Pre-arranged	117	213		97.9
Department				<0.001	
	Accident and emergency	0	4		1.2
	Burns	0	2		0.6
	Cardiac	0	2		0.6
	Chest surgery	5	11		4.7
	Chinese medicine	4	2		1.8
	Dermatology	0	8		2.4
	Elderly care	2	0		0.6
	Gastrointestinal	6	16		6.5
	General surgery	35	44		23.4
	Gynaecology	0	22		6.5
	Haematology	0	4		1.2
	Incretion	1	0		0.3
	Midwifery	0	2		0.6
	Neurosurgery/neurology	2	6		2.4
	Operating theatres	2	6		2.4
	Orthopaedics	4	16		5.9
	Otolaryngology	4	2		1.8
	Ophthalmology	11	25		10.7
	Paediatrics/neonatal	2	4		1.8
	Respiratory	30	29		17.5
	Stomatology	12	8		5.9
	Urology	4	2		1.8

[†] Chi-square test.

3.2. Principal Component Analysis

An exploratory factor analysis was carried out by performing a principal component analysis (PCA) with an orthogonal varimax rotation for the 16 individual items at a significance level of $p < 0.001$. Orthogonal varimax rotation is chosen because of the unrelated nature of produced factors [38]. Factor solution was based on Bartlett's test showing a significant correlation between items (Chi-square = 2444.295; $p < 0.001$) and the Kaiser–Meyer–Olkin test for sample adequacy measuring 0.838 which is considered 'great' by Field [39]. These indices implied that the matrix was well suited for factor analysis. An initial analysis was run to obtain eigenvalues for each component in the data. Five summated indices from the 16 question items that had eigenvalues greater than 1.0 represented five different scales. Factor 1 consisted of three items accounting for 34.7% of the variance, Factor 2 represented four items accounting for 14.7% of the variance, and Factor 3 had four items which accounted for an additional 8.5% of the variance. Factors 4 and 5 had three and two items which accounted for 6.8% and 6.4% of the variance, respectively. The total variance is 71.2%. Given the large sample size and the convergence of the scree plot and Kaiser's criterion on five components, this is the number of components that were retained in the final analysis. Table 3 shows the factor loadings after rotation. These five scales of design were identified as sensory, facilities, spatial, lighting and thermal, and seating design.

Table 2. Descriptive analysis.

Questionnaire Items	Response * (%)					Mean	SD
	1	2	3	4	5		
Cleanliness	0.0	0.0	3.6	37.7	58.8	4.55	0.565
Air freshness	0.0	0.0	5.9	35.3	58.8	4.53	0.607
Noise	0.0	2.1	13.9	34.1	49.9	4.32	0.789
A thermally comfortable environment	0.0	0.6	12.8	53.1	33.5	4.20	0.671
Seating sufficiency [†]	0.3	0.6	12.8	53.7	32.6	4.18	0.689
Adequate illumination [‡]	0.0	1.5	23.4	45.4	29.7	4.03	0.769
Spaciousness	1.2	1.5	23.1	52.2	22.0	3.92	0.783
Availability of daylight	0.0	1.5	28.2	50.7	19.6	3.88	0.725
Seating comfort	0.9	4.5	29.1	38.6	27.0	3.86	0.896
Architectural design of the space	0.6	5.3	40.1	34.4	19.6	3.67	0.870
Pleasant colour scheme	1.2	7.7	40.9	38.6	11.6	3.52	0.842
Indoor plants, interior/exterior landscaping	2.7	8.0	42.7	39.5	7.1	3.40	0.840
Exterior view	2.4	11.3	50.4	30.0	5.9	3.26	0.825
Presence of coordinated art objects	3.9	16.9	44.5	26.4	8.3	3.18	0.943
Furniture layouts	3.9	8.9	57.3	26.4	3.6	3.17	0.789
Entertainment facilities	1.2	21.1	49.3	23.4	5.0	3.10	0.828

* 1: Least important; 2: Unimportant; 3: Neither important nor unimportant; 4: Important; 5: Most important.
[†] Adequate number of seats. [‡] Overall lighting: artificial and natural lighting combined.

Table 3. Rotated component matrix of questionnaire items.

Questionnaire Items	Components				
	Sensory	Facilities	Spatial	Lighting and Thermal	Seating
Air freshness	0.856	-	-	-	-
Cleanliness	0.833	-	-	-	-
Noise	0.719	-	-	-	-
Exterior view	-	0.805	-	-	-
Presence of coordinated art objects	-	0.781	-	-	-
Indoor plants, interior/exterior landscaping	-	0.696	-	-	-
Entertainment facilities	-	0.574	-	-	-
Furniture layouts	-	-	0.791	-	-
Architectural design of the space	-	-	0.755	-	-
Pleasant colour scheme	-	-	0.669	-	-
Spaciousness	-	-	0.566	-	-
Availability of daylight	-	-	-	0.792	-
Adequate illumination	-	-	-	0.720	-
A thermally comfortable environment	-	-	-	0.574	-
Seating sufficiency	-	-	-	-	0.805
Seating comfort	-	-	-	-	0.773
Cronbach's alpha coefficient (0.870)	0.792	0.768	0.784	0.850	0.714
Percentage of explained variance (71.2)	34.714	14.713	8.482	6.819	6.437

3.3. Internal Consistency Reliability

The reliability of each attribute was examined by the Cronbach's alpha coefficients. The obtained values of the reliability estimates were all greater than 0.70 as shown in Table 3, indicating a strong internal reliability among items with the same attributes. Table 3 also shows the internal consistency reliability level (Cronbach's alpha coefficients) for each generated factor that 0.792 for sensory design, 0.768 for facility design, 0.784 for spatial design, 0.850 for lighting and thermal design and 0.714 for seating design. Combined, these five factors explained 71.2% of all variables and the Cronbach's alpha coefficient for the overall scale was 0.870.

3.4. Relationship of Personal Information and Perspectives of Design Factors

Non-parametric tests were carried out on 16 questionnaire items, as shown in Table 4. Results show that there is a significant difference in perspectives between male and female outpatients in the sensory design aspect including air freshness, cleanliness and noise. Age has a significant effect on the perspectives of both sensory and seating design aspects. Patients do not have significantly different perspectives regarding the appointment type. However, the findings suggest the number of visits to the hospital has influenced their perspectives on spatial and seating aspects, which represent six out of sixteen items in the whole questionnaire.

Table 4. Comparison of mean principal component analysis (PCA) scores between demographic variables.

Questionnaire Items		Components				
		Sensory	Facilities	Spatial	Lighting and Thermal	Seating
Gender	Male	4.31(0.53)	3.22(0.64)	3.56(0.66)	3.89(0.68)	4.01(0.69)
	Female	4.45(0.52)	3.24(0.67)	3.57(0.63)	3.99(0.72)	4.01(0.71)
	p-value †	**0.046** *	0.703	0.929	0.184	0.952
Age(year)	18–25	4.34(0.55)	3.30(0.61)	3.59(0.67)	3.95(0.72)	4.14(0.77)
	26–35	4.45(0.51)	3.30(0.69)	3.66(0.64)	4.04(0.65)	4.12(0.67)
	36–50	4.42(0.52)	3.16(0.53)	3.47(0.54)	3.92(0.72)	3.87(0.64)
	>50	4.19(0.56)	3.00(0.91)	3.37(0.81)	3.69(0.82)	3.69(0.77)
	p-value ‡	**0.002** *	0.169	0.265	0.839	**0.007** *
Appointment type	Emergency	4.57(0.35)	3.54(0.70)	4.03(0.47)	4.57(0.53)	4.36(0.56)
	Pre-arranged	4.39(0.53)	3.23(0.66)	3.56(0.64)	3.95(0.70)	4.01(0.71)
	p-value †	0.978	0.329	0.071	0.109	0.562
Number of visit	1–2	4.51(0.51)	3.31(0.66)	3.64(0.67)	4.11(0.75)	4.14(0.69)
	3–4	4.39(0.52)	3.31(0.64)	3.70(0.62)	4.02(0.68)	4.11(0.71)
	5–10	4.30(0.64)	3.10(0.79)	3.38(0.59)	3.85(0.79)	3.86(0.76)
	>10	4.36(0.43)	3.16(0.55)	3.45(0.62)	3.79(0.55)	3.89(0.64)
	p-value ‡	0.143	0.774	**0.008** *	0.755	**0.010** *

† Mann–Whitney *U*-test; ‡ Kruskal–Wallis test; * $p < 0.05$.

4. Discussion

Among the dimensions of the waiting environment evaluated by 337 outpatients, *cleanliness* (mean = 4.55) was ranked as the most important indicator, followed by 'air freshness' (mean = 4.53) and 'noise' (mean = 4.32). 'Entertainment facilities' (mean = 3.10) was the least important indicator in the overall waiting environment, which together with 'furniture layouts' (mean = 3.17) and the 'presence of coordinated art objects' (mean = 3.18) were ranked as the bottom three (Table 2). The reason for relatively low scores in these three items may be due to the physical situation in both surveyed hospitals. On the one hand, there is a big number of outpatients every day (average number of daily hospital outpatient visits was nearly 1500 in the surveyed departments), and the waiting rooms are always full of patients and their families, some patients even have to wait outside in the corridor. All the patients are waiting to meet care providers in a queue, not like in some hospitals in developed countries with an electronic queuing system to display patient numbers on a flat screen. The outpatients in the surveyed hospital have to pay more attention to being called rather than entertain themselves. On the other hand, some outpatients suffered from illness and had no mood to watch TV or reading newspapers at all. Airflow rate has an important to role to play in ventilation [40] and the perception of air freshness. However, although most outpatients did not consider entertainment facility and art objects in hospital as important as other aspects, they are welcomed in some inpatient unit design [41] and suggestions have been made to supply newspaper or magazines to improve the entertainment in particular departments [42].

Results also show that the overall rating scores are quite high ranging from 3.10 to 4.55, indicating the importance of questionnaire items. Six out of sixteen items had mean scores higher than 4 (=important) and the remaining ten items all had mean scores higher than 3 (=neither important nor unimportant). Regarding constructed dimensions, sensory design, seating design, the design of lighting and thermal environments was of concern to the respondents since all the eight surveyed items under these dimensions had the highest mean scores, compared with items under the dimensions, 'spatial' and 'facilities'.

From the results of surveyed items, relatively high rating scores indicate respondents prefer more natural daylight and adequate illumination when they are waiting for the doctors. A large body of evidence shows that exposure to bright artificial light and daylight is effective in reducing depression and improving patients' mood [43]. Furthermore, research indicates the exposure to light is critical to patient and staff health and wellbeing in healthcare settings [44–46]. However, excessive daylight can also cause visual discomfort through glare and distraction, which is affected by window design. A big window size could let more daylight come in and at the same time will consume more energy in heating or cooling [47]. Therefore, there is a trade-off that needs to be reconciled for designing the window area and providing enough daylight in the room [48].

Mean scores received from female outpatients were higher than male in most of the surveyed items except the architectural design of the space (mean scores 3.61 vs. 3.78); indoor plants, interior/exterior landscaping (3.39 vs. 3.43) and seating comfort (3.85 vs. 3.89). Results from non-parametric test show there is a significant difference of perspectives on sensory design aspects between male and female. Female respondents highly evaluated the importance of air freshness (4.55) and cleanliness (4.60). It is a fairly natural response because these two items are frequently reported in the literature as most important attributes of a physical environment. Also, women in China are more responsible for housing and cleaning than men, which may lead to a higher expectation of the environment they spend hours staying. Cleanliness is also considered the most important as it was ranked the first place in the mean scores of respondents' perspectives. Such result is in line with another study conducted by the authors in which cleanliness was ranked in first place with regards to the hospital accommodation environment by a group of surveyed inpatients [17] and care providers [49]. Similar results were also found by Shah and Dickinson [50], who investigated the factors patients might consider when choosing hospitals and the weight of the factors during decision making. The results from their study showed hospital cleanliness was the most important factor followed by hospital reputation and seven other factors. For patients, cleanliness is inexorably related with healthcare associated infections (HAIs; therefore, it is necessary for any healthcare facility to maintain a high standard of cleanliness.

Noise is the most frequently studied environmental factor in hospitals that relates to both patient and care providers [43]. Hagerman et al. [51] found a relationship between the noise level in patient rooms and patient satisfaction. They also found a bad acoustics environment is likely to produce a bad working environment for staff that could adversely affect the patients. Males and females have different perspectives on the ambient environment because males may be more tolerant than females [52]. This argument is supported by this study that females consider noise is more important in the hospital design than their male counterparts (4.40 vs. 4.18).

The analysis also shows that females are more perceptive than men on the summated five factors except they have the same mean score on seating environment (4.01 vs. 4.01). A significant difference in perspectives based on gender was found for sensory design within the constructed dimensions. Females considered that sensory design (air freshness, cleanliness and noise) to be more important (mean score = 4.45) than males (mean score = 4.15). This result suggests that women are more perceptive of overall sense-sensitive design factors, which is in accordance with previous research showing women have greater sensitivity in sensory factors than men [53–55].

There is a significant difference in respondents' perspectives based on age for the dimensions of sensory and seating design. In this study, seating dimensions include two indicators: seating sufficiency and seating comfort. Results show that the younger respondents thought seating dimension more important than older respondents, where mean score from 18–25 years old outpatients was 4.14 and 3.69 by outpatients >50. It is speculated that younger respondents require more interaction in the waiting room rather than merely waiting for the doctor's call. Evidence has been highlighted in one of Ulrich's [1] paper that in waiting rooms, day rooms, and lounges, the widespread practice of arranging seating side-by-side along the walls of a room markedly inhibits social interaction among patients or other users, which corroborates long-held views by Holahan [56] and Sommer and Ross [57]. Younger outpatients also evaluated all the five design dimensions with higher mean scores compared with older outpatients (>50 years). However, lateral comparison within the five dimensions indicates that older patients thought sensory design factor more important (mean score = 4.19) and the facilities design factor (mean score = 3.00) the least important.

Most research has assessed patients' satisfaction as the patient outcome measure through evaluation of healthcare service and quality of care. Very few studies link the number of patient visits to how their satisfaction with the healthcare environment. This study has identified outpatients' perspectives regarding their frequency of visits to the hospital. Respondents who have been to the hospital for more than five times have relatively low mean scores (lower than 4.00) in all four dimensions except the 'sensory design'. This may be because people who visited the hospital more times will have fewer expectations of their known environments. People are more perceptive of environments with which they are unfamiliar. It may also relate to hospital waiting times in China; patients who are more familiar with the environment would choose to visit at a time which is less crowded. It is also reflected in the answers from the interview that some outpatients "prefer to come in the afternoon to avoid waiting and delay in the morning".

In addition, other than the sensory design factor, seating design has been rated more important than the other three environmental aspects. Significant differences in outpatients' perspectives were found in the spatial and seating dimensions. Patients visiting hospital less frequently thought the seating environment more important than patients having visited hospital more often. This result is agreed by other researchers; for example, Tsai et al. [58] found that the 'body-contact environment', including seating environment, is perceived less favourable by first-time visitors. This may be due to their dissatisfaction with the high volume of patients and insufficient seats. In China, a similar situation is shared as they have the largest number of outpatients in hospitals every day. As discussed earlier, good arrangement of seats may enhance the interaction between patients. Nevertheless, the waiting room's crowded conditions often lead to patients' discomfort with their surroundings. Therefore, such factors make them more important in outpatients' perspectives and deserving of more attention in the design process.

5. Research Limitations

This study entails several limitations. First, this study excluded respondents who are younger than 18 years old. The overall response rates reached 84.25%. Unlike other studies, this response rate excludes questionnaires with missing values; it would be possible to have more valid responses to certain questions if missing values were included.

Second, although respondents' social and demographic information was obtained, there is more information worth recording from outpatients, such as educational background and monthly income. However, considering the cultural preferences and circumstances where the questionnaire survey was conducted, some patients may feel the answers to questions on income are too private to give.

Third, due to the unbalanced development of healthcare in urban and rural communities in China, there are differences in urban and rural healthcare infrastructures. This study focused on outpatients' perspectives from two urban healthcare centres, and the findings may not be representative of the overall Chinese healthcare facilities.

Fourth, the relatively high response rate in the present study promised a good interpretation of the results, as response rates are crucial concerning the generalisability of results [59]. However, it is necessary to point out that differences may exist in perception between non-respondents (uncompleted surveys) and respondents. To reduce the influence of the lacking responses, the present study was completed anonymously to diminish the influence of social desirability, gratitude and dependence, therefore, it is feasible not to include responses from incomplete surveys.

Finally, validation is a continuous process, and further studies are required to confirm these results. The experimental nature of these studies may have included bias in questionnaire responses. Thus, there is a need to replicate findings using confirmatory statistical methods using the data from non-experimental, routine studies.

6. Conclusions

Many studies have explored outpatients' satisfaction regarding the healthcare service they receive from specific dimensions, such as waiting experience, interaction with care providers and quality of care. However, findings from such research seldom provide useful insights on not-so-tangible aspects of healthcare design in the decision making. This research was aimed to address the need for a reliable and valid instrument associated with design indicators of waiting areas in healthcare facilities via assessing outpatient's perspectives.

The present questionnaire is a 16-item self-completed questionnaire on a five-point Likert-type scale. Questionnaire development was based on an extensive literature review and the views of sample outpatients who felt that the relevant aspects of outpatients' perspectives were adequately covered. The developed questionnaire is acceptable to outpatients while maintaining comprehensibility in its coverage of important aspects of patient experience in outpatient departments [30]. Descriptive and principal component analyses were conducted on the obtained data; non-parametric tests were applied to identify if there were significant differences in patients' perspectives of the constructed PCA factors with demographic variables. A relatively good response rate and minor comments reported by the participants indicate that this questionnaire can be used to understand and extract outpatients' perspectives of the importance of design indicators on the healthcare waiting environment. The instrument has undergone a testing process for reliability and validity, which supports its application as a measure of patients' perspectives. The core scales are supported by the results of the factor analysis. PCA confirmed the hypothesised dimensional structure of the questionnaire, yielding five factors. The initial grouping of the items as shown in this study should be considered in relation to the explorative nature of the research. The interpretation of the factors was based on the loadings of each item on each factor. Items with the highest loadings on a factor were considered as most strongly related to that factor and thus referred to that factor [60]. The high levels of internal consistency reliability for information and hospital standards suggest that the items comprising these hypothesised scales are sufficiently related.

Among the investigated design indicators, 'design for cleanliness' was ranked as the most important, followed by 'air freshness' and 'noise', both with mean scores above 4.30, indicating that they are high on the agenda for inpatients. These three indicators formed 'sensory' design in the constructed dimensions. In other words, respondents considered conventional environmental design factors to be highly important, more than other design factors. The lowest ranked item was 'entertainment facilities', followed by 'furniture layouts' and 'presence of coordinated art objects'. All three had mean scores above 3.10 and were part of the 'facilities design' factor, indicating that, although the factors were at the bottom of the list, the respondents considered them to be important, but not as important as the environmental design factor.

The research findings are important for integrating outpatients' perspectives in the design process. It is interesting to anticipate the integration of evidence-based design of healthcare facilities with perspectives of facility users. However, further research is required to validate and confirm current findings in different geographic regions.

Acknowledgments: The research reported in this article is part of the first author's Ph.D. dissertation, available at: https://dspace.lboro.ac.uk/2134/12621. The research was conducted when both authors were based at Loughborough University. The study was funded by the UK Engineering and Physical Sciences Research Council (EPSRC) via the Health and Care Infrastructure Research and Innovation Centre (HaCIRIC) (grant reference: EP/D039614/1).

Author Contributions: Y.Z. and M.M. conceived and designed the experiments; Y.Z. collected the data; and Y.Z. and M.M. analysed the data and wrote the paper.

Conflicts of Interest: The authors declare no conflict of interest.

References

1. Ulrich, R.S. Effects of healthcare environmental design on medical outcomes. In *World Congress on Design and Health*; Dilani, A., Ed.; International Academy for Design and Health: Stockholm, Sweden, 2001; pp. 49–59.

2. Schweitzer, M.; Gilpin, L.; Frampton, S. Healing Spaces: Elements of Environmental Design That Make an Impact on Health. *J. Altern. Complement. Med.* **2004**, *10*, S71–S83. [CrossRef] [PubMed]

3. Huisman, E.R.C.M.; Morales, E.; van Hoof, J.; Kort, H.S.M. Healing environment: A review of the impact of physical environmental factors on users. *Build. Environ.* **2012**, *58*, 70–80. [CrossRef]

4. Devlin, A.S.; Arneill, A.B. Health Care Environments and Patient Outcomes. *Environ. Behav.* **2003**, *35*, 665–694. [CrossRef]

5. Douglas, C.H.; Douglas, M.R. Patient-centred improvements in health-care built environments: Perspectives and design indicators. *Health Expect.* **2005**, *8*, 264–276. [CrossRef] [PubMed]

6. Mourshed, M. Interoperability Based Optimisation of Architectural Design. Ph.D. Thesis, National University of Ireland, Cork, Ireland, 2006.

7. Lawson, B.; Phiri, M.; Wells-Thorpe, J. *The Architectural Healthcare Environment and Its Effects on Patient Health Outcomes*; NHS Estates: London, UK, 2003.

8. Reiling, J. Safe design of healthcare facilities. *Qual. Saf. Health Care* **2006**, *15*, i34–i40. [CrossRef] [PubMed]

9. Fottler, M.D.; Ford, R.C.; Roberts, V.; Ford, E.W. Creating a healing environment: The importance of the service setting in the new consumer-oriented healthcare system. *J. Healthc. Manag.* **2000**, *45*, 91–106. [PubMed]

10. Zhao, Y.; Mourshed, M.; Wright, J.A. Factors influencing the design of spatial layouts in healthcare buildings. In *Proceedings of the 25th ARCOM Conference, Nottingham, UK, 7–9 September 2009*; Dainty, A.R.J., Ed.; Association of Researchers in Construction Management (ARCOM): Reading, UK, 2009.

11. Gutteling, J.J.; de Man, R.A.; Busschbach, J.J.V.; Darlington, A.-S.E. Quality of health care and patient satisfaction in liver disease: The development and preliminary results of the QUOTE-Liver questionnaire. *BMC Gastroenterol.* **2008**, *8*, 25. [CrossRef] [PubMed]

12. Woodward, C.A.; Ostbye, T.; Craighead, J.; Gold, G.; Wenghofer, E.F. Patient satisfaction as an indicator of quality care in independent health facilities: Developing and assessing a tool to enhance public accountability. *Ame. J. Med. Qual.* **2000**, *15*, 94–105. [CrossRef] [PubMed]

13. Woodring, S.; Polomano, R.C.; Haagen, B.F.; Haack, M.M.; Nunn, R.R.; Miller, G.L.; Zarefoss, M.A.; Tan, T.L. Development and Testing of Patient Satisfaction Measure for Inpatient Psychiatry Care. *J. Nurs. Care Qual.* **2004**, *19*, 137–148. [CrossRef] [PubMed]

14. Schulmeister, L.; Quiett, K.; Mayer, K. Quality of life, quality of care, and patient satisfaction: Perceptions of patients undergoing outpatient autologous stem cell transplantation. *Oncol. Nurs. Forum* **2005**, *32*, 57–67. [CrossRef] [PubMed]

15. Daniel, O. Perception and Patient Satisfaction: A Case Study of Olabisi Onabanjo University Teaching Hospital Sagamu, Nigeria. Master's Thesis, Blekinge Institute of Technology, Karlskrona, Sweden, 2009.

16. Eriksen, L. Measuring Patient Satisfaction with Nursing Care: A Magnitude Estimation Approach. In *Measurement of Nursing Outcomes*, 2nd ed.; Springer: New York, NY, USA, 2001.

17. Zhao, Y.; Mourshed, M. Design indicators for better accommodation environments in hospitals: Inpatients' perceptions. *Intell. Build. Int.* **2012**, *4*, 199–215. [CrossRef]

18. Zeithaml, V.A.; Bitner, M.J.; Gremler, D.D. *Services Marketing*, 6th ed.; McGraw-Hill Education: New York, NY, USA, 2012.

19. Bergenmar, M.; Nylén, U.; Lidbrink, E.; Bergh, J.; Brandberg, Y. Improvements in patient satisfaction at an outpatient clinic for patients with breast cancer. *Acta Oncol.* **2006**, *45*, 550–558. [CrossRef] [PubMed]

20. Singh, H.; Haqq, E.D.; Mustapha, N. Patients' perception and satisfaction with health care professionals at primary care facilities in Trinidad and Tobago. *Bull. World Health Organ.* **1999**, *77*, 356–360. [PubMed]

21. Gremigni, P.; Sommaruga, M.; Peltenburg, M. Validation of the Health Care Communication Questionnaire (HCCQ) to measure outpatients' experience of communication with hospital staff. *Patient Educ. Couns.* **2008**, *71*, 57–64. [CrossRef] [PubMed]

22. Laine, C.; Davidoff, F.; Lewis, C.E.; Nelson, E.C.; Nelson, E.; Kessler, R.C.; Delbanco, T.L. Important elements of outpatient care: A comparison of patients' and physicians' opinions. *Ann. Intern. Med.* **1996**, *125*, 640–645. [CrossRef] [PubMed]

23. Wiggins, M.N.; Coker, K.; Hicks, E.K. Patient perceptions of professionalism: Implications for residency education. *Med. Educ.* **2009**, *43*, 28–33. [CrossRef] [PubMed]

24. Jha, V.; Bekker, H.L.; Duffy, S.R.G.; Roberts, T.E. Perceptions of professionalism in medicine: A qualitative study. *Med. Educ.* **2006**, *40*, 1027–1036. [CrossRef] [PubMed]

25. Boscarino, J.A. Patients' perception of quality hospital care and hospital occupancy: Are there biases associated with assessing quality care based on patients' perceptions? *Int. J. Qual. Health Care* **1996**, *8*, 467–477. [CrossRef] [PubMed]

26. Lurie, J.D.; Berven, S.H.; Gibson-Chambers, J.; Tosteson, T.; Tosteson, A.; Hu, S.S.; Weinstein, J.N. Patient preferences and expectations for care: Determinants in patients with lumbar intervertebral disc herniation. *Spine* **2008**, *33*, 2663–2668. [CrossRef] [PubMed]

27. Brennan, P.F.; Strombom, I. Improving health care by understanding patient preferences: The role of computer technology. *J. Am. Med. Inform. Assoc.* **1998**, *5*, 257–262. [CrossRef] [PubMed]

28. Gittell, J.H.; Fairfield, K.M.; Bierbaum, B.; Head, W.; Jackson, R.; Kelly, M.; Laskin, R.; Lipson, S.; Siliski, J.; Thornhill, T.; et al. Impact of relational coordination on quality of care, postoperative pain and functioning, and length of stay: A nine-hospital study of surgical patients. *Med. Care* **2000**, *38*, 807–819. [CrossRef] [PubMed]

29. Baughman, C.; Spurling, L.; Mangoni, A.A. Provision of warfarin education to hospital inpatients. *Br. J. Clin. Pharmacol.* **2008**, *66*, 416–417. [CrossRef] [PubMed]

30. Garratt, A.M.; Bjaertnes, Ø.A.; Krogstad, U.; Gulbrandsen, P. The OutPatient Experiences Questionnaire (OPEQ): Data quality, reliability, and validity in patients attending 52 Norwegian hospitals. *Qual. Saf. Health Care* **2005**, *14*, 433–437. [CrossRef] [PubMed]

31. Singer, S.; Götze, H.; Möbius, C.; Witzigmann, H.; Kortmann, R.D.; Lehmann, A.; Höckel, M.; Schwarz, R.; Hauss, J. Quality of care and emotional support from the inpatient cancer patient's perspective. *Langenbecks Arch. Surg.* **2009**, *394*, 723–731. [CrossRef] [PubMed]

32. Chandra, A.; Finlay, J.B.; Paul, D.P. Overall outpatient satisfaction and its components: Perceived changes at the Huntington VA Medical Center over five years. *Hosp. Top.* **2006**, *84*, 33–36. [CrossRef] [PubMed]

33. Shyu, Y.-I.L.; Tang, W.-R.; Tsai, W.-C.; Liang, J.; Chen, M.-C. Emotional support levels can predict physical functioning and health related quality of life among elderly Taiwanese with hip fractures. *Osteoporos. Int.* **2006**, *17*, 501–506. [CrossRef] [PubMed]

34. Donabedian, A. Evaluating the Quality of Medical Care. *Milbank Q.* **2005**, *83*, 691–729. [CrossRef] [PubMed]

35. Cronbach, L.J. Coefficient alpha and the internal structure of tests. *Psychometrika* **1951**, *16*, 297–334. [CrossRef]

36. Nunnally, J.C.; Bernstein, I.H. *Psychometric Theory*, 3rd ed.; McGraw-Hill: New York, NY, USA, 1994.

37. Bland, J.M.; Altman, D.G. Multiple significance tests: The Bonferroni method. *BMJ* **1995**, *310*, 170. [CrossRef] [PubMed]

38. Costello, A.B.; Osborne, J.W. Best practices in exploratory factor analysis: Four recommendations for getting the most from your analysis. *Pract. Assess. Res. Eval.* **2005**, *10*, 1–9.

39. Field, A. *Discovering Statistics Using IBM SPSS Statistics*, 4th ed.; Sage Publications Ltd.: London, UK, 2013.

40. Lydon, G.P.; Ingham, D.B.; Mourshed, M.M. Ultra clean ventilation system performance relating to airborne infections in operating theatres using CFD modelling. *Build. Simul.* **2014**, *7*, 277–287. [CrossRef]

41. Dobrohotoff, J.T.; Llewellyn-Jones, R.H. Psychogeriatric inpatient unit design: A literature review. *Int. Psychogeriatr.* **2011**, *23*, 174–189. [CrossRef] [PubMed]

42. Walsh, M.; Knott, J.C. Satisfaction with the emergency department environment decreases with length of stay. *Emerg. Med. J.* **2010**, *27*, 821–828. [CrossRef] [PubMed]

43. Ulrich, R.S.; Zimring, C.; Zhu, X.; DuBose, J.; Seo, H.B.; Choi, Y.S.; Quan, X.; Joseph, A. A review of the research literature on evidence-based healthcare design. *Health Environ. Res. Des. J.* **2008**, *1*, 61–125. [CrossRef]

44. Campbell, S.S.; Kripke, D.F.; Gillin, J.C.; Hrubovcak, J.C. Exposure to light in healthy elderly subjects and Alzheimer's patients. *Physiol. Behav.* **1988**, *42*, 141–144. [CrossRef]

45. Shikder, S.; Mourshed, M.; Price, A.D.F. Therapeutic lighting design for the elderly: A review. *Perspect. Public Health* **2012**, *132*, 282–291. [CrossRef] [PubMed]

46. Lockley, S.W.; Barger, L.K.; Ayas, N.T.; Rothschild, J.M.; Czeisler, C.A.; Landrigan, C.P.; Harvard Work Hours; Health and Safety Group. Effects of health care provider work hours and sleep deprivation on safety and performance. *Jt. Comm. J. Qual. Patient saf.* **2007**, *33*, 7–18. [CrossRef]

47. Shikder, S.H.; Mourshed, M.; Price, A.D.F. Optimisation of a daylight-window: Hospital patient room as a test case. In *Proceedings of the International Conference on Computing in Civil and Building Engineering, Nottingham, UK, 30 June–2 July*; Tizani, W., Ed.; Nottingham University Press: Nottingham, UK, 2010.

48. Mourshed, M.; Kelliher, D.; Keane, M. Optimised building form for environmental sustainability. In *Proceedings of the Conference on Global Built Environment: Towards an Integrated Approach for Sustainability, Preston, UK, 11–12 September 2006*; Mourshed, M., Ed.; Global Built Environment Network: Preston, UK, 2006.

49. Mourshed, M.; Zhao, Y. Healthcare providers' perception of design factors related to physical environments in hospitals. *J. Environ. Psychol.* **2012**, *32*, 362–370. [CrossRef]

50. Shah, J.; Dickinson, C.L. Establishing which factors patients value when selecting urology outpatient care. *Br. J. Med. Surg. Urol.* **2010**, *3*, 25–29. [CrossRef]

51. Hagerman, I.; Rasmanis, G.; Blomkvist, V.; Ulrich, R.; Eriksen, C.A.; Theorell, T. Influence of intensive coronary care acoustics on the quality of care and physiological state of patients. *Int. J. Cardiol.* **2005**, *98*, 267–270. [CrossRef] [PubMed]

52. Yu, L.; Kang, J. Effects of social, demographical and behavioral factors on the sound level evaluation in urban open spaces. *J. Acoust. Soc. Am.* **2008**, *123*, 772–783. [CrossRef] [PubMed]

53. Velle, W. Sex differences in sensory functions. *Perspect. Biol. Med.* **1987**, *30*, 490–522. [CrossRef] [PubMed]

54. Feine, J.S.; Bushnell, M.C.; Miron, D.; Duncan, G.H. Sex differences in the perception of noxious heat stimuli. *Pain* **1991**, *44*, 255–262. [CrossRef]

55. Fillingim, R.B.; Maixner, W. Gender differences in the responses to noxious stimuli. *Pain Forum* **1995**, *4*, 209–221. [CrossRef]

56. Holahan, C. Seating patterns and patient behavior in an experimental dayroom. *J. Abnorm. Psychol.* **1972**, *80*, 115–124. [CrossRef] [PubMed]

57. Sommer, R.; Ross, H. Social Interaction on a Geriatrics Ward. *Int. J. Soc. Psychiatry* **1958**, *4*, 128–133. [CrossRef]

58. Tsai, C.-Y.; Wang, M.-C.; Liao, W.-T.; Lu, J.-H.; Sun, P.-H.; Lin, B.Y.-J.; Breen, G.-M. Hospital outpatient perceptions of the physical environment of waiting areas: The role of patient characteristics on atmospherics in one academic medical center. *BMC Health Serv. Res.* **2007**, *7*, 198. [CrossRef] [PubMed]

59. Sitzia, J.; Wood, N. Patient satisfaction with cancer chemotherapy nursing: A review of the literature. *Int. J. Nurs. Stud.* **1998**, *35*, 1–12. [CrossRef]

60. Capra, M.G. Factor Analysis of Card Sort Data: An Alternative to Hierarchical Cluster Analysis. *Proc. Hum. Factors Ergon. Soc. Annu. Meet.* **2005**, *49*, 691–695. [CrossRef]

© 2017 by the authors. Licensee MDPI, Basel, Switzerland. This article is an open access article distributed under the terms and conditions of the Creative Commons Attribution (CC BY) license (http://creativecommons.org/licenses/by/4.0/).

Article

The Usability Study of a Proposed Environmental Experience Design Framework for Active Ageing

Masa Noguchi [1,*], Nan Ma [2], Catherine Mei Min Woo [1], Hing-wah Chau [1] and Jin Zhou [1]

1   ZEMCH Lab, Faculty of Architecture, Building and Planning, The University of Melbourne,
    Melbourne, VIC 3010, Australia; catherine.woo@unimelb.edu.au (C.M.M.W.); chauh@unimelb.edu.au (H.C.);
    jin.zhou@unimelb.edu.au (J.Z.)
2   Department of Architecture, School of Design, University of Pennsylvania, Philadelphia, PA 19104, USA;
    nanma1@design.upenn.edu
*   Correspondence: masa.noguchi@unimelb.edu.au; Tel.: +61-3-9035-8193

Received: 14 August 2018; Accepted: 20 November 2018; Published: 28 November 2018

Abstract: Growing ageing population today may be necessitating building design decision makers to reconsider the indoor environmental quality (IEQ) standards in a way that accommodates senior occupants' diverse and individual needs and demands. An experience design approach to rationalising and individualising end-user experience on how to utilise tangible products may serve to reflect user perceptions. Generally, architectural design practices tend to incorporate neither IEQ monitoring and analysis data, nor environmental experience design today. In response to the need for filling this gap, the authors of this paper conducted a feasibility study previously that led to structuring and defining an 'Environmental Experience Design' (EXD) research framework. Based on the previous case study on the collective spatial analysis and IEQ monitoring results, this paper further explored the usability and applicability of this proposed EXD framework particularly to the previously documented aged care facility in Victoria, Australia, which has been stressing active ageing agendas. This EXD framework usability experiment helped to build the capacity for engaging the subjectivity and objectivity of end users' expectations, desires, and requirements in the architectural design thinking process. Nonetheless, due to the limitation of this initial and fundamental usability study's resources and the objective, the necessity of adjusting the scale and scope of EXD analyses emerged. Moreover, the universality of this EXD research framework usage under various architectural typologies and user conditions yet require further attempts and investigations.

Keywords: architectural design thinking; user-centric building design; environmental experience design; residential aged care facilities; design for active ageing

1. Introduction

The population of Australia is ageing [1–3]. There were 3.5 million senior citizens who were aged 65 years and over in 2014 taking up 15% of the population [4]. It is estimated that the proportion of senior citizens will rise to 26% in 2051 and to 27% in 2101 [5–7]. The population of Victoria follows the tendencies shown in the wider Australian population (Table 1). As of September 2017, the estimated Victorian population was 6,179,249 [8]. This is an increase of almost 23% since June 2005 [9]. Residents aged 55 and over cover nearly a quarter of the population (22.4%) and those aged 65 years and above form 16.7% of Victoria's population [1]. The proportion of the population aged 65 years and over is expected to go up to 17.4% in 2021, 18.8% in 2031, 20.4% in 2041, and 21.8% in 2051 [9]. The greatest proportional shift in next few decades to be expected is the number of Victorians aged 85 years and above is projected to increase from 2.6% of the population in 2017 to 4.6% in 2051 [9]. There were 27% of the population aged 65 and over born in a non-English speaking country in contrast to 20% born domestically [10].

Table 1. Population of people aged 55 years and over in Victoria [1].

Age	Men	Women	Total
55+	336,334 (11.0%)	354,377 (11.3%)	690,711 (22.4%)
65+	251,532 (8.2%)	265,111 (8.5%)	516,643 (16.7%)
75+	130,624 (4.3%)	154,595 (5.0%)	285,219 (9.2%)
85+	47,602 (1.6%)	79,750 (2.6%)	127,352 (4.1%)
All ages	3,056,434	3,122,815	6,179,249

The rise of senior population in Victoria led to the increase of aged care facility establishments and the architectural design may need to serve as an agent of engagement for societal needs. The design decisions today tend to be made without favouring user experiences and this challenge might be derived from the discrepancies between prescribed building codes and user perception. The role of built environments may become more prominent in managing increasing sensitives and vulnerabilities that come with ageing. As the Australian population ages, the state of Victoria is actively working towards facilitating effective spatial design strategies through an integrated framework for "active ageing" (Figure 1) [11–18].

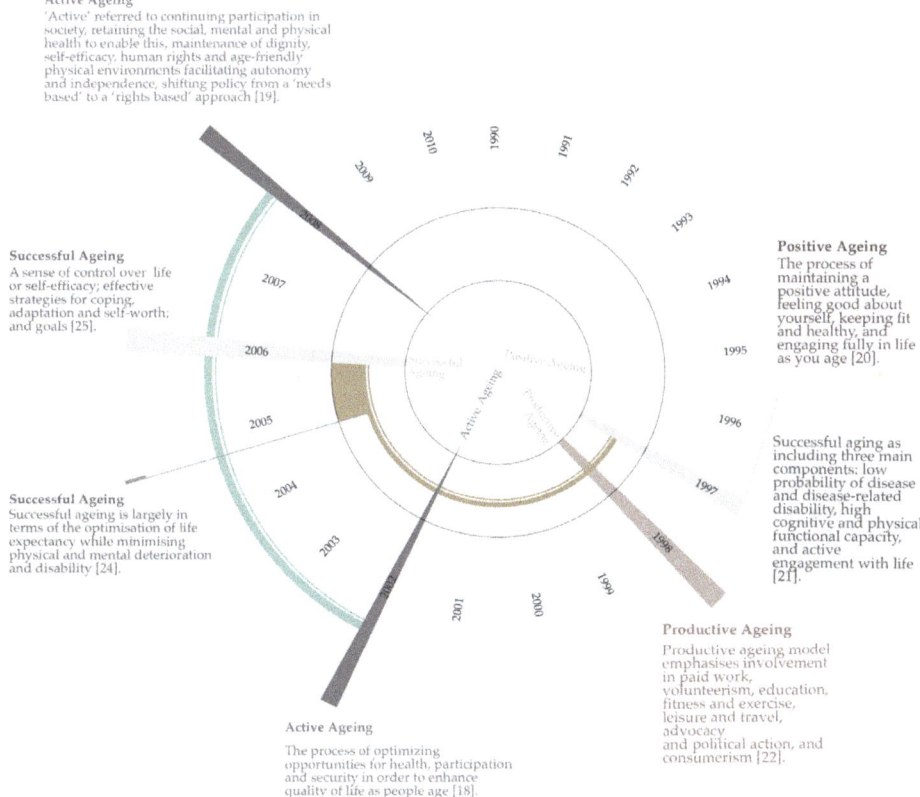

Figure 1. Aged care agenda timeline.

It is worth noting that the "experience design" has already been applied to industrial designs and the "user experience design" focuses on improving products' interface to facilitate the usage in response to end-users' diverse physical and psychological needs and demands [19]. Pine and Gilmore (1999) stress that experience designs encompass both passive and active participation of end-users [20]. Architectural design has an impact on users' physical and perceived comfort levels in the built environment. Nonetheless, the notion of such experience design is barely applied to architectural design practices today.

Data analysis or 'programming' facilitates stakeholders' design orientation in response to the project objectives identified for the achievement [21]. De Giuli et al. (2012) articulates the significant impacts of indoor environmental quality (IEQ) research around thermal, acoustic, and visual conditions, as well as air quality on occupants' health and wellbeing in the built environment [22]. IEQ data collected and/or simulated may provide some insights or indications of what the building in question can perform to maintain the acceptable levels of the occupants' health and wellbeing [23]. Therefore, IEQ standards need to be well incorporated into architectural design decision making; nonetheless, it may be worth noting that these physical indicators alone need not reflect occupants' perceived quality. Neither does the environmental design data itself serve as a direct architectural design decision-making tool.

This study explores this challenge through implementing an 'Environmental Experience Design' (EXD) research framework, which was previously proposed by this paper's authors who reviewed the related theories that help to illustrate human physical and psychological needs and demands [24]. The proposed EXD framework devises "function analysis" techniques that help to categorise occupants' requirements, desires and expectations in the built environment [25]. This study tested the proposed EXD research framework as a systematic approach to further identifying relevant design solutions towards activating senior citizens for the improvement of their health and wellbeing.

2. Environmental Experience Design Research Framework Review

A human-environment integrated approach that assists architectural design stakeholders in understanding the occupants' physical and psychological needs and demands is required. In response to this need, the authors of this paper proposed a conceptual 'Environmental Experience Design' (EXD) research framework in 2017 [24]. This proposed EXD framework was designed to identify overall project objectives, analyse user perception, and propose design strategies and solutions. It is an interdisciplinary trajectory that is relatively new to architectural practices, aimed at embracing a human-environment integration into the design decision making process. The EXD framework devises a function analysis methodology that helps to identify "performance of a user function" and refine the design procedure to "fulfil a user requirement" by questioning what user needs are and how designers meet them [25]. The Function Analysis System Technique (FAST) diagram is first generated as a process to logically visualise the project's key objectives or functions for prioritisation. The FAST diagram serves as a map or pathway towards the scrutiny of possible design solutions in response to the users' physical and psychological needs and demands identified (Figure 2).

Contextualising the human-environment relationship is of importance in the built environment, since the space affects users' activities of daily living [26,27]. Spatial design strategies need to be set in a way that matches both objective physical parameters (e.g., natural and built environment settings) and subjective user perception (e.g., psychological needs and demands). In the EXD research framework, in response to FAST implementation results, user experience related functions and the associated spatial design strategies and solutions are contextualised through the development of a human-environment matching 'EXD evaluation matrix' (Figure 2) [24]. The enumeration helps to visualise the relationship between the occupants' physical and psychological requirements, desires and expectations, and the potential architectural design strategies and solutions applied to shaping the space accordingly.

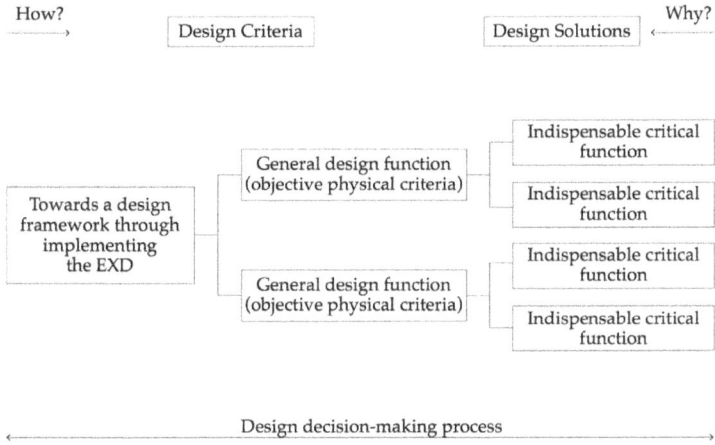

Figure 2. General Function Analysis System Technique (FAST) diagram image.

The EXD research framework was developed conceptually as a tool that any stakeholders can apply to identifying projects' objectives and relevant design solutions (Figure 3). Nonetheless, the actual usability and applicability are still in question. Thus, the following sections will demonstrate how this conceptual EXD tool can be applied to upgrading a selected aged care facility in Victoria, Australia, in consideration of the Victorian government's active ageing agenda.

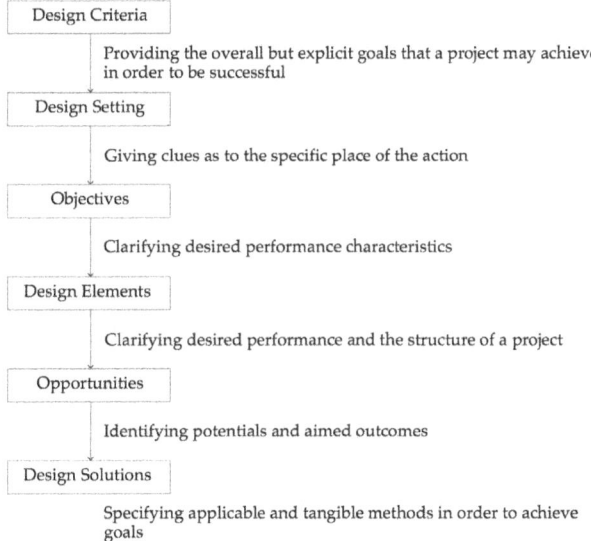

Figure 3. Environmental Experience Design (EXD) matrix evaluation process for human-environment element integration.

3. Proposed EXD Framework Implementation

This study revisited an aged care facility located in Victoria, Australia, with the aim to test the usability and applicability of the Environmental Experience Design (EXD) research framework proposed initially by authors of this paper in 2017 (Figure 4) [24]. It is also aimed at demonstrating the human-environment matching mechanism oriented towards activating the senior residents for their health and wellbeing.

Figure 4. Exterior view of Adare SRS studied.

To make sure that the selected aged care facility was designed to maintain the minimal levels of physical indoor environmental quality (IEQ) conditions, IEQ monitoring of the temperature, and the levels of particulate matter and carbon dioxide concentration was conducted over a one-week period from 29th May to 4th June 2017 (Figure 5) [24]. The study confirmed that the overall IEQ conditions were generally satisfactory. Nonetheless, it also led to stressing a potential consideration, as indicated "Although the thermal condition of both bedroom and communal space are in the lower range of thermal comfort zone defined by ANSI/ASHRAE Standard 55-2013, a warmer environment is recommended ... the 20–24 °C comfort zone is not warm enough for older adults and older adults generally prefer a warmer environment than younger subjects" (Figure 6) [24].

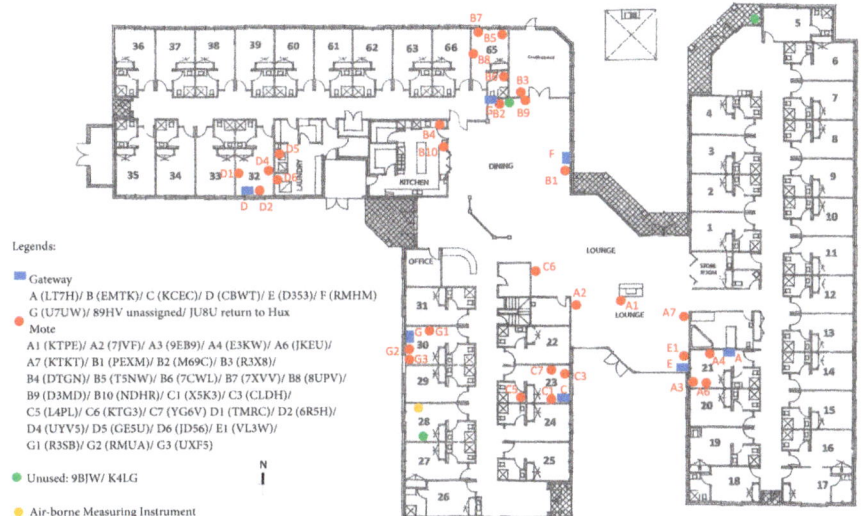

Figure 5. Locations of Indoor Environmental Quality (IEQ) Measurements.

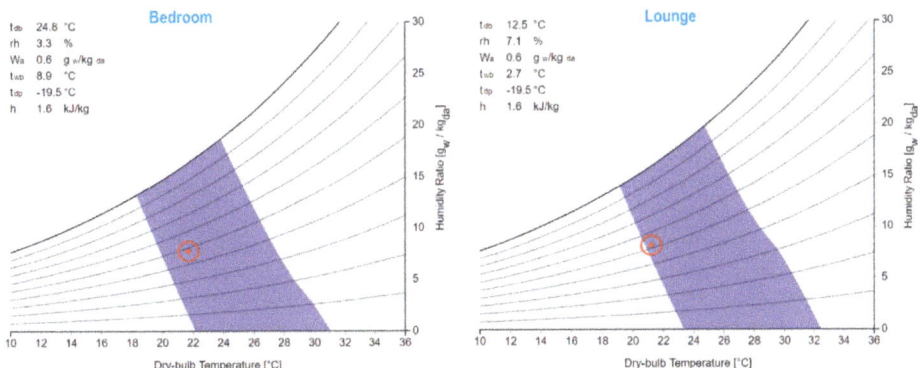

Figure 6. Thermal condition of the bedroom (**left**) and lounge space (**right**) in psychometric chart. The shaded areas represent comfort zone boundary [24].

Based on the spatial analysis of the aged care facility revisited, the EXD FAST diagram was developed with the aim to outline the relevant functions that reflect the senior residents' general needs and demands around their activities of daily living, as well as to identify sensitive spaces and the upgrading approaches to activating the elderly physically and psychologically. The function analysis aims to encompass both the subjectivity and objectivity of users' needs and demands; therefore, all stakeholders including not only the residents and visitors but also designers and builders may ideally be involved in the thinking process. Nonetheless, due to the main aim of this study that attempts to demonstrate and analyse the framework usability, the EXD FAST diagram was shaped by the researchers who observed, documented, and analysed the building, in addition to the IEQ monitoring on behalf of the stakeholders (Figure 7).

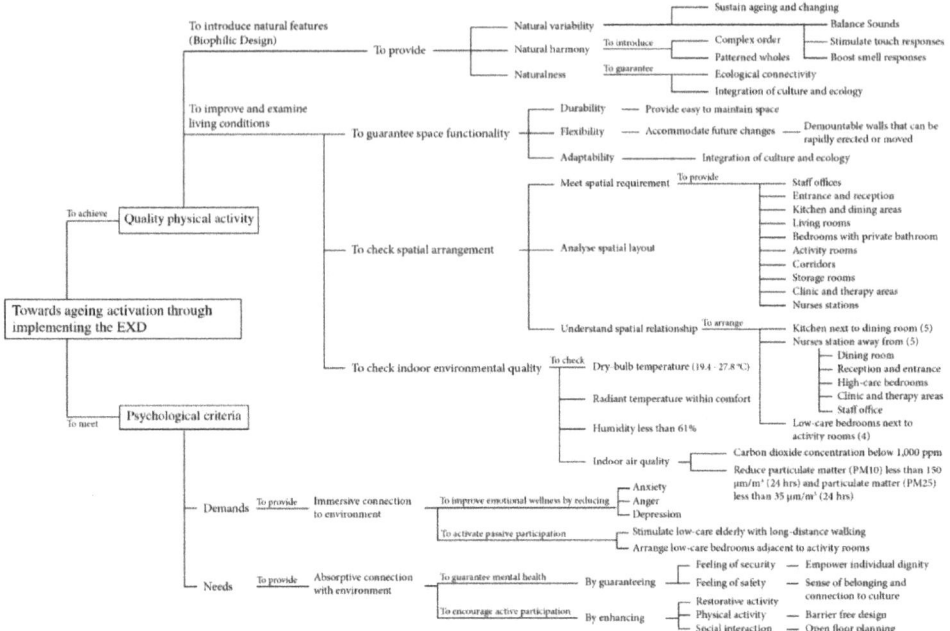

Figure 7. EXD FAST diagram of senior residents' physical and metaphysical functions.

Based on the observations of the spatial settings throughout the ground floor, level differences are well avoided, and this inclusive design fundamental allows for enhanced accessibility and safety of the senior users with mobility aids. Partitions set between the dining and kitchen area tend to discourage visual and physical interaction, while limiting the circulation of unfavourable smells. Staff areas, including the office, reception, and nurses' station are located next to the dining room, which serves as the residents' social activities. The bedrooms were dimensionally adequate and were lit by natural light coming through the transparent window that is centered in the external wall. Outdoor spaces are equipped with accessible garden pedestrian paths, communal spaces with barbeque facilities, visually stimulating artefacts, and seasonal plants—those that to some extent contribute to the creation of atmosphere that supports the notion of active aging. Based on the spatial setting observation and FAST diagram results, an EXD evaluation matrix was developed with the aim of identifying the potential upgrading solutions that reflect the Victorian government's active ageing agenda (Table 2).

81

Table 2. EXD evaluation matrix enumerating potential senior users' physical attributes and psychological perceptions towards identification of possible design solutions [24].

Design Criteria	Design Settings	Objectives	Design Elements	Opportunities	Design Solutions
	Natural setting	Natural variability	Sounds	Providing pleasing effects	Euphonic sounds increase
				Providing soothing effects	Chaotic sounds reduction
			Touch responses	Interaction with vegetation	Garden walking pathways
				Interaction with animals	Freedom of feeding pets
			Smell responses	Olfactory stimuli	Different scented plants
			Ageing and changing	The rhythm of life	Seasonal plants assembly
		Natural harmony	Complex order	Rich sensory information	Different types of flowers and trees gardening
			Patterned wholes	Intriguing balance between boring and overwhelming	Symmetric and fractal geometries of integrated pavements
		Naturalness	Outdoor gardens	Ecological connectivity	Communal spaces in the garden
			Garden crafts	Culture and ecology integration	The elderly engaged gardening
		Spatial adaptability	Present spatial needs	Achieving the elderly's satisfactions	Calming colours of wall and ceiling paintings
					Natural textured material of floorings
					Two layered curtains: gauze and fabric curtains with natural colours
					Freedom of bringing personal belongings
					Lockable interior doors
					Keep gardens open
		Spatial flexibility	Future spatial needs	Pliable temporal limitation	Moveable partitions
		Spatial durability	Ease of maintenance spaces	Cleanliness guarantee	Flooring with vinyl
Physical Activation Criteria	Built environment setting	Spatial arrangement	Spatial requirement	Friendly, comfortable and welcoming living conditions	Kitchen adheres to dining areas
					Communicative living rooms
					Bedrooms with private bathroom
					Multi-activity rooms
					Corridors with hand rails
					Clinic and therapy areas
			Spatial relationship		A nurse's station
					Kitchen next to dining room (5)
					Nurses station away from dining room (5)
					Nurses station away from reception and entrance (5)
					Nurses station away from high-care residents' bedrooms (5)
					Nurses station away from clinic and therapy areas (5)
					Nurses station away from staff office (3)
					Low-care elderly's bedrooms next to activity rooms (4)

Table 2. *Cont.*

Design Criteria	Design Settings	Objectives	Design Elements	Opportunities	Design Solutions
		Indoor environmental quality	Spatial layout		Using bubble diagrams
			Dry bulb temperature		Direct heat exposure avoidance: ranging between 19.4 to 27.8 °C
					Shading provision: within the comfort zone
			Radiant temperature	Elderly wellbeing, comfort and health	Fresh air ventilation: less than 61%
			Relative humidity		Fresh air ventilation: Carbon Dioxide concentration below 1000 ppm
			Indoor air quality		Particulate matter (PM10) less than 150 $\mu m/m^3$ (in 24 hrs)
					Particulate matter (PM25) less than 35 $\mu m/m^3$ (in 24 hrs)
	Psychological demands	Emotional wellness	Anxiety reduction	Harmful effects on health and the overall quality of life	Respect the elderly's personal choices and decisions
			Anger reduction		
			Depression reduction		
		Passive participation	Activating physical walking	Increase engagement and participation	Low-care elderly's bedrooms with a distance away to living rooms
			Visual connection to activity rooms		Layout of the low-care elderly's bedrooms adjacent to activity rooms
		Security	Empowering a sense of individual dignity	Being free from injury	Non-hierarchical spaces
					Manageable spaces
					Controllable spaces
Psychological Activation Criteria	Psychological needs	Safety	Generating a sense of belongingness	Being protected from risk	Locating residential care facilities in the neighbourhood
					Rooms have visual connections with outdoors
					Stimulating the elderly bringing distinctive vernacular objects
					Stimulating the elderly engaged gardening
					A space with the passage of time
					Level difference avoidance
					Slip-resistant and firm flooring surfaces
		Active participation	Barrier-free spatial conditions	Physical activity stimulation	Physical and visual barriers elimination
					Wide interior doors, corridors and turning spaces associated with the elderly movements
			Open floor planning	Social interaction	Maximising the use of limited spaces
					Minimising partitions
			Multi-activity rooms' design	Restorative activity	Evoking and developing the elderly's interests

In this study, the EXD evaluation matrix contents were further simplified and illustrated to articulate the selected aged care facility's human-environment relationship (Figure 8). EXD visualisation serves as a medium that all stakeholders, such as the facility's staff, users, visitors, and appointed architects, can understand today's spatial circumstances and the future upgrading potentials that contribute to turning passive senior residents into active ones for their physical and mental health and wellbeing.

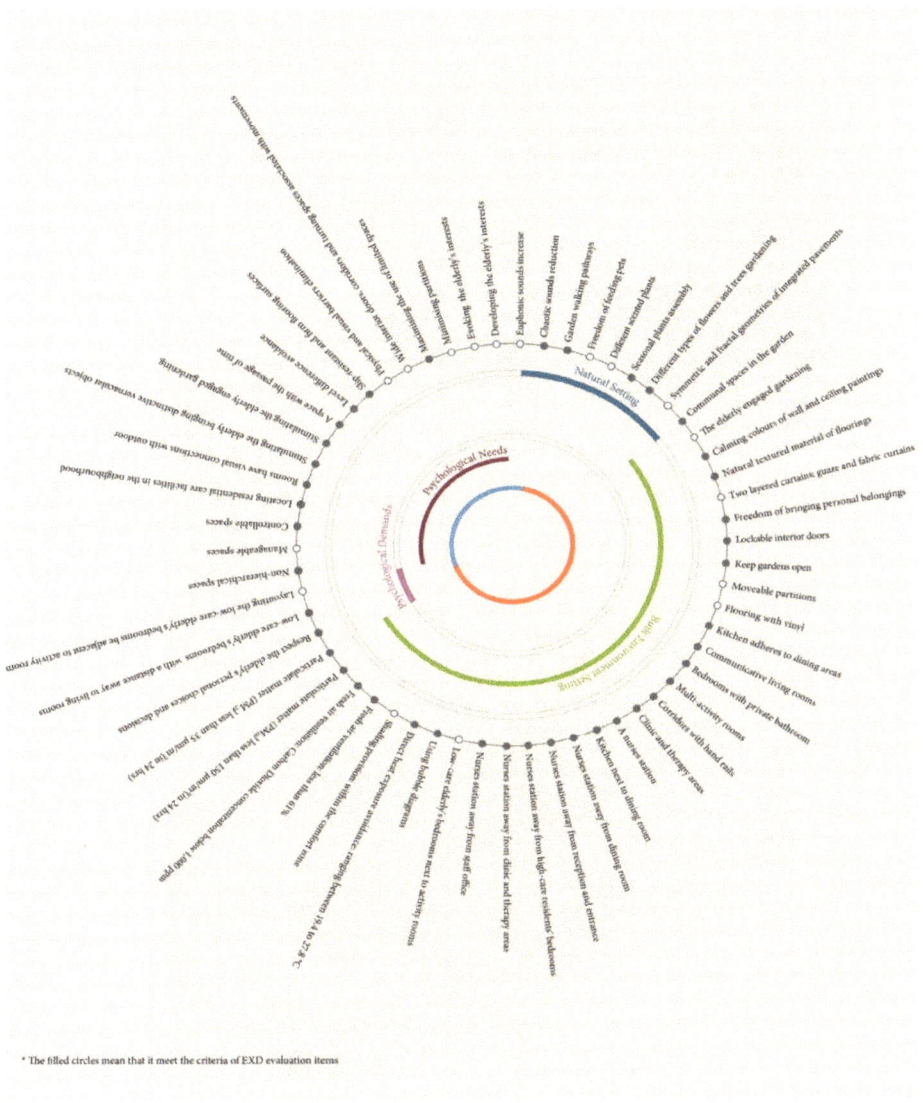

* The filled circles mean that it meet the criteria of EXD evaluation items

Figure 8. EXD human-environment relationship visualisation.

EXD Spatial Upgrading Potentials

The EXD evaluation matrix was contextualised for value visualisation and it led to identifying some potential architectural design or space upgrading solutions for active ageing (Figure 9). Three spaces of the selected aged care facility were used for the demonstration of the proposed EXD framework. Before-and-after design upgrades of these selected spaces were illustrated based on the EXD evaluation outcomes. The annotations appearing in Figures 9–11 reflect the design solutions that are listed on the 6th column from the left of Table 2.

Figure 9. An existing bedroom unit (**top**) and the EXD upgrading potentials (**bottom**).

Suggestions for the minor renovation included: the replacement of fabric carpets with anti-bacterial titles for the elderly users' smooth walk and enhanced sanitation; the change of existing window blinds to double-layered fabric curtains, which allow for more flexibility in modulating the intensity of natural light being introduced into the internal space; and, the introduction of indoor potted plants that encourage the senior residents' engagement with nature in the controlled built environment. These renovations can be realised within the building's existing structure, dimension, volume, and layout; nonetheless, some low-care units that are occupied by immobile senior users may desire major high-care upgrades (Figure 10).

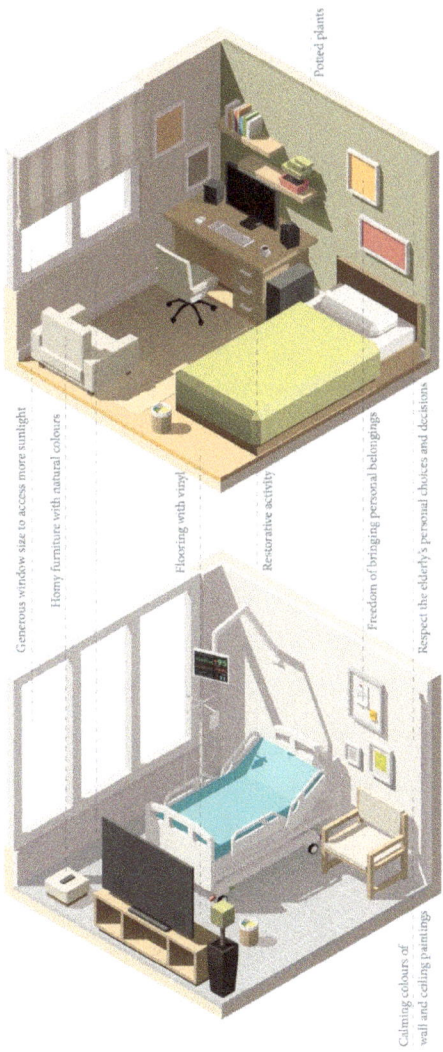

Figure 10. An existing low-care room (**top**) and a suggested high-care upgrade (**bottom**).

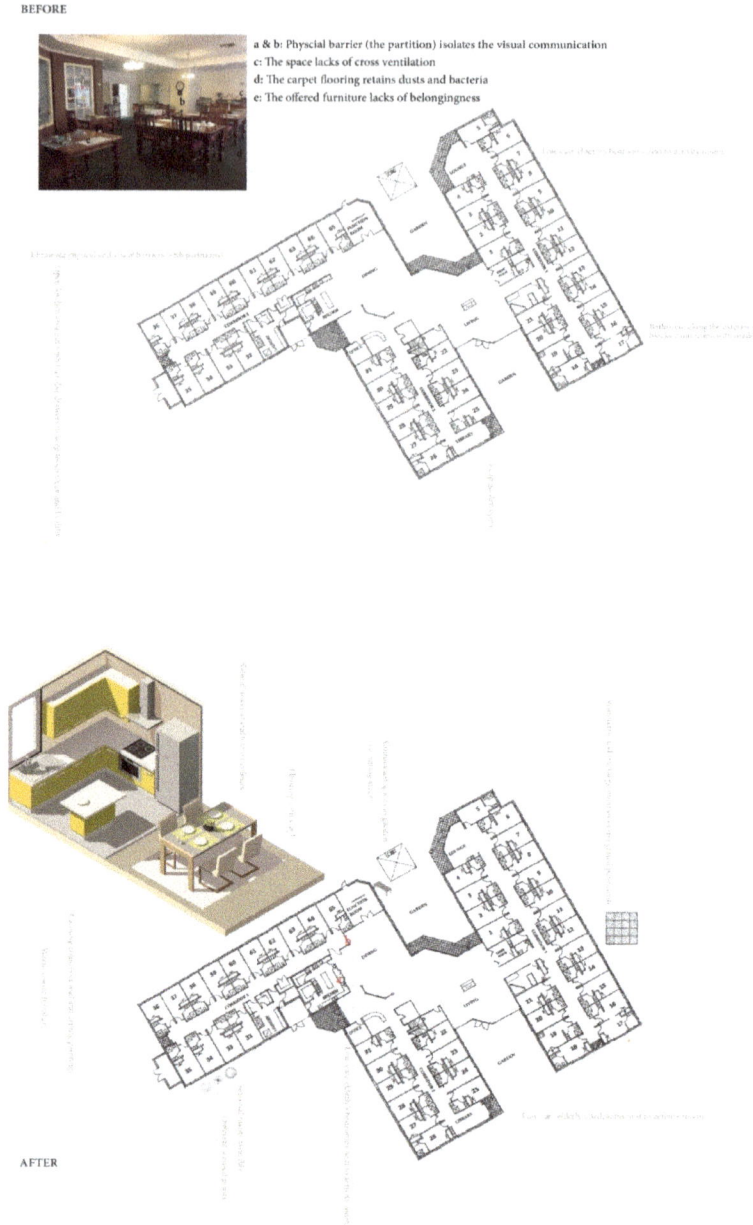

Figure 11. Proposed articulation for visual and physical connectivity between spaces.

Suggestions for the ground floor improvement extended to the relocation of low care bedrooms to places adjacent to rooms that embrace social and physical activities (Figure 11). This may create opportunities for immobile senior residents to be linked to visual and acoustic stimuli. The reduction of physical and visual barriers (i.e., open floor planning) might also lead to the enhancement of users' social networking opportunities. Non-structural rigid partitions can be replaced with collapsible or movable partitions that can separate or open spaces according to the users' specific needs and demands.

Outdoor settings may also need to be designed for enhancement of the senior occupants' safety, accessibility, comfort, and stimulation for active ageing. Exposure to the full spectrum of natural light may contribute to activating the occupants along with humans' circadian rhythms. The proposed designs included features that aim to promote visual, acoustic, and olfactory stimuli, and the placement of vegetation (and pets) would also serve as an active ageing driver. Dubos (1980) argues that "people want to experience the sensory, emotional, and spiritual satisfactions that can be obtained only from an intimate interplay, indeed from an identification with the places which [they] live" [28]. However, such human satisfaction may be continuously reliant on perceiving and responding to sensory variability [29]. Senses of sound, touch, and smell form the sensory richness [29]. Although walking pathways exist in the selected aged care facility, there is no intent or consideration of adding pleasant visual and olfactory stimuli to the users' experience. If these spaces are equipped with interactive green gardens, senior residents may be more attracted to taking a stroll or even the upkeep.

4. Conclusions

The proposed 'Environmental Experience Design' (EXD) research framework was an attempt to connect the domains of environmental design and experience design. It addressed complexities of the human-environment relationship and served as a design decision-making support tool that helps visualise end-users' needs and demands, as well as build a pathway towards identification of the associated design solutions. This study was an extension of the authors' previous research project that conceptualised the EXD research framework itself. In this paper, the usability and application to a selected aged care facility's spatial design upgrades for active ageing were explored and demonstrated conceptually. In addition to the IEQ building performance check, this EXD framework usability experiment helped to build the capacity for engaging the subjectivity and objectivity of end users' expectations, desires, and requirements in the architectural design thinking process. Nonetheless, this study was limited to an initial and fundamental demonstration of the EXD decision making process that targeted the potential design improvements of a selected aged care facility in Victoria, Australia; therefore, the universality of this EXD research framework usage under various architectural typologies and user conditions still requires further attempts and investigations. Moreover, the scale and scope of EXD analyses need to be narrowed and focused much further for in-depth exploration of each functional space in the built environment, while the stakeholders' direct involvement in the function analysis stage is necessitated to identify their precise perceived needs and demands rather than the speculations. A subsequent validation study of the EXD experiential effect on IEQ improvements may require further justification of the usability and universality.

Author Contributions: M.N. and N.M. led overall research activities and contributed to structuring the proposed environmental experience design diagnostic framework and writing this paper. C.M.M.W. contributed to the editorial coordination. H.-w.C. documented the aged care facility selected in Victoria. J.Z. contributed to the IEQ monitoring and data analysis.

Acknowledgments: The authors would like to express their sincere gratitude to Adare SRS which allowed them to access their aged care facility rooms for in-depth documentation and IEQ monitoring.

Conflicts of Interest: The authors declare no conflict of interest.

References

1. Australian Bureau of Statistics. *Australian Demographic Statistics, March 2017*; Australian Bureau of Statistics: Canberra, Australia, 2017.

2. Australian Bureau of Statistics. *Population by Age and Sex, Victoria*; Australian Bureau of Statistics: Canberra, Australia, 2006.

3. Australian Institute of Health and Welfare. *Australia's Welfare 2011*; Australian Institute of Health and Welfare: Canberra, Australia, 2011.

4. Australian Institute of Health and Welfare. *Australia's Welfare 2015*; Australian Institute of Health and Welfare: Canberra, Australia, 2015.

5. Australian Bureau of Statistics. *Population by Age and Sex, Australian States and Territories*; Australian Bureau of Statistics: Canberra, Australia, 2011.

6. Australian Bureau of Statistics. *Health of Older People in Australia: A Snapshot, 2004–2005*; Australian Bureau of Statistics: Canberra, Australia, 2006.

7. Australian Institute of Health and Welfare. *Australia's Health 2010*; Australian Institute of Health and Welfare: Canberra, Australia, 2010.

8. Australian Bureau of Statistics. *Population by Age and Sex, Victoria*; Australian Bureau of Statistics: Canberra, Australia, 2016.

9. Australian Bureau of Statistics. *National Health Survey—First Results: Australia 2014–2015*; Australian Bureau of Statistics: Canberra, Australia, 2015.

10. Federation of Ethnic Communities' Councils of Australia. *Review of Australian Research on Older People from Culturally and Linguistically Diverse Backgrounds*; Federation of Ethnic Communities' Councils of Australia: Canberra, Australia, 2015.

11. World Health Organisation. Active Ageing: A Policy Framework. 2002. Available online: http://apps.who.int/iris/bitstream/10665/67215/1/WHO_NPH_02.8.pdf (accessed on 26 September 2017).

12. Bowling, A. Enhancing Later Life: How Older People Perceive Active Ageing? *Aging Ment. Health* **2008**, *12*, 293–301. [CrossRef] [PubMed]

13. Kendig, H.; Browning, C. Positive Ageing: Facts and Opportunities. *Med. J. Aust.* **1997**, *167*, 409–441. [PubMed]

14. Rowe, J.W.; Kahn, R.L. Successful Ageing. *Gerontologist* **1997**, *37*, 433–440. [CrossRef] [PubMed]

15. Kerschner, H.; Pegues, J.A.M. Productive Aging: A Quality of Life Agenda. *J. Am. Dietet. Assoc.* **1998**, *98*, 1445–1448. [CrossRef]

16. Kalache, A.; Gatti, A. Active ageing: A policy framework. *Adv. Gerontol.* **2003**, *11*, 7–18. [PubMed]

17. Bowling, A.; Dieppe, P. What Is Successful Ageing and Who Should Define It? *Br. Med. J.* **2005**, *331*, 1548–1551. [CrossRef] [PubMed]

18. Bowling, A.; Iliffe, S. Which Model of Successful Ageing Should Be Used? Baseline Findings from a British Longitudinal Survey of Ageing. *Age Ageing* **2006**, *35*, 607–614. [CrossRef] [PubMed]

19. Hassenzahl, M. *Experience Design: Technology for All the Right Reasons*; Morgan and Claypool Publishers: Philadelphia, PA, USA, 2010.

20. Pine, B.J.; Gilmore, J.H. *The Experience Economy: Work Is Theatre & Every Business a Stage*; Harvard Business School Press: Cambridge, MA, USA, 1999.

21. Farbstein, J.; Wener, R.; McCunn, L.J. *Planning the Built Environment: Programming. Research Methods for Environmental Psychology*; Wiley: Malaysia, 2016.

22. De Giuli, V.; da Pos, O.; de Carli, M. Indoor Environmental Quality and Pupil Perception in Italian Primary Schools. *Build. Environ.* **2012**, *56*, 335–345. [CrossRef]

23. Al horr, Y.; Arif, M.; Katafygiotou, M.; Mazroei, A.; Kaushik, A.; Elsarrag, E. Impact of Indoor Environmental Quality on Occupant Well-being and Comfort: A Review of the Literature. *Int. J. Sustain. Built Environ.* **2016**, *5*, 1–11. [CrossRef]

24. Ma, N.; Chau, H.; Zhou, J.; Noguchi, M. Structuring the Environmental Experience Design Research Framework through Selected Aged Care Facility Data Analyses in Victoria. *Sustainability* **2017**, *9*, 12. [CrossRef]

25. Dell'Isola, A.J. *Value Engineering: Practical Applications: for Design, Construction, Maintenance & Operations*; R.S. Means: Kingston, NY, USA, 1997.

26. Horgas, A.L.; Wilms, H.U.; Baltes, M.M. Daily Life in Very Old Age: Everyday Activities as Expression of Successful Living. *Gerontologist* **1998**, *38*, 556–568. [CrossRef] [PubMed]

27. Krause, N. Neighborhood Deterioration, Religious Coping, and Changes in Health during Late Life. *Gerontologist* **1998**, *38*, 653–664. [CrossRef] [PubMed]

28. Dubos, R. *The Wooing of Earth*; Scribner: New York, NY, USA, 1980.
29. Kellert, S. *Kinship to Mastery: Biophilia in Human Evolution and Development*; Island Press: Washington, DC, USA, 1997.

© 2018 by the authors. Licensee MDPI, Basel, Switzerland. This article is an open access article distributed under the terms and conditions of the Creative Commons Attribution (CC BY) license (http://creativecommons.org/licenses/by/4.0/).

Article

The Impact of Outdoor Views on Students' Seat Preference in Learning Environments

Zhonghua Gou [1,*], Maryam Khoshbakht [1] and Behnam Mahdoudi [2]

[1] School of Engineering and Built Environment, Griffith University, Gold Coast, QLD 4215, Australia; m.kh@griffith.edu.au

[2] Department of Architecture, Kish International Campus, University of Tehran, Tehran 1417466191, Iran; Behnam.mahdoudi@alumni.ut.ac.ir

* Correspondence: z.gou@griffith.edu.au or gouzhonghua@gmail.com; Tel.: +61-7-5552-9510

Received: 10 June 2018; Accepted: 24 July 2018; Published: 28 July 2018

Abstract: A Good learning environment should support students' choices and attract them to stay. Focusing on outdoor views, this research explores two questions: How important outdoor views are in seat selection in learning environments? How do the view elements influence students' seating behaviors in learning environments? A seat preference survey and view elements and occupancy rate measurements were conducted in a university library building in Gold Coast, Australia. This study not only echoes the previous research indicating that territory and privacy are important factors for choosing seats in a learning environment; more importantly, this study contributes to the literature with evidence that outdoor views might be an important factor for seat preference. Specifically, sky views and shading views were found positively related to occupancy rate. Based on this point, open views with appropriate shading were found as an optimal outdoor view composition. The singularity of greenery views would less likely be attractive to building occupants.

Keywords: outdoor views; learning environment; seat preference; sky; shading; greenery

1. Introduction

The design of a learning environment has a significant impact on students' attitudes, behaviors and achievements [1,2]. A successful learning environment should attract and encourage students to stay. Therefore, seat preference is one of the important performance criteria for research that seeks to understand suitably designed learning environments. There are many design factors influencing seat preference. Haghighi and Jusan [3] surveyed 370 public high school students in Iran using a questionnaire comprised five-point Likert-type scales evaluating classroom architectural items and students' achievement motivation. The study categorized three types of physical settings in learning environments that might affect students' seat selections and related behaviors: fixed settings (such as windows, walls, doors, and generally the outline of elements that shape the learning space), semi-fixed settings (such as radiators, bookshelves, and bulletin boards) and flexible settings (such as chairs and desks that can be moved). Yildirim Cagatay and Ayalp [4] assessed the effects of three different colors (cream, blue and pink) on the interior walls using a semantic differential scale composed of 11 bipolar adjectives for students in a High School in Turkey. The results highlighted the importance of wall color on the perception of classrooms and found that the use of different colors in interior spaces of a classroom had a statistically significant effect on the perceptual performance of the male students. Other studies [5–8] investigated diverse environmental aspects, such as daylighting, sunlight, room acoustics, temperature, which would affect students' perception and preference. Among all potential environmental factors, territoriality is addressed as the most important factor explaining seat preferences in different types of learning environments [9]. In environmental psychology, territoriality refers to how people use space to communicate occupancy of areas or possessions [10]. In practice, learning space arrangements with well-defined areas can have a positive influence on students' social interaction as well as on task behaviors [9,11].

Although previous studies successfully identified a series of environmental design factors influencing seat selection and related behaviors, outdoor views are largely missing in those studies [8,12]. Many green building standards reward building design that provides occupants with a connection between indoors and outdoors through the introduction of views into the regularly occupied areas of the building. For example, LEED (Leadership in Energy and Environmental Design) allocates up to four points to Quality Views [13]. There are many benefits for doing so, such as reinforcing human circadian rhythms and enhancing psychological wellbeing [14–16]. There is a long research track record showing the benefit of outdoor views [17–19]. Among all, views containing natural elements such as trees and sky are highlighted in psychological studies as preferred views [20,21]. Although outdoor views have many benefits to building users, there are few studies correlating views with seat preference or duration of stay.

In sum, the study of stay and seat preference of learning environments is mainly focused on interior elements such as territory, colors and the like, while outdoor elements that would influence occupants' preferences, such as views, are largely missing. The research of linking views to seat preference or duration of stay is of great importance for designing learning environments. In learning environments such as a library, students have the choice to select their own seats. A good learning environment is supposed to support their choices and to attract their stays. The research aims to explore two important research questions missing in the literature:

How important outdoor views are in seat selection in learning environments?

How do the view elements influence students' seating behaviors in learning environments?

2. Methodology

2.1. The Surveyed Building

To explore the research questions, a library building was selected for the study. The building is Griffith University Gold Coast Library (Figure 1), located at a coastal city in South East Queensland on the east coast of Australia (coordinates: 27.962522 S and 153.379988 E). Gold Coast experiences a humid subtropical climate with warm winters (June–August) and hot, humid summers (December to February). The library building under study is a redevelopment of an old library building with an extension to accommodate the growing book collections and to provide more learning spaces. The project completed in 2012. Most workstations in this library are located next to windows, directly facing the outside to maximize accessibility to views. The middle space is dedicated to bookshelves. This layout design considers the fact that library users or students preferred seats near windows or daylit areas [22]. This library building and its view-oriented layout provide a unique opportunity for investigating the relationship between views, daylight peformance, and occupants behaviours. In total, 88 workstations at level 3 were selected for this study (Figure 2). They all are located in the periphery of the library space, facing the outside.

A lighting measurement was conducted to measure the illuminance at the selected workstations using Minolta T-10 (range: 0.01 to 299,99lx; accuracy ±1 digit). The measurement was conducted on two days in the summer of Southern Hemisphere: 1 December 2015 (Sunny Day) and 2 December 2015 (Cloudy Day) as well as two days in the winter of Southern Hemisphere: 6 July (Sunny Day) and 11 July 2016 (Cloudy Day). Figure 3 combines the four days' data and compares four orientations using box plots. The north-facing workstations had a higher average desktop illuminance. This condition is different from the northern hemisphere where south-facing space is supposed to receive the most daylight during the day. The west-facing workstations were shaded by louvers; therefore, the desktop illuminance was lower than others. The desktop illuminance in east-facing workstations fell into a larger spectrum than others, especially in the morning. The south-facing workstations tended to have more outliers on desktop illuminance. The outliers came from the workstations with large sky views.

During the measurement, no direct sunlight incidence was observed in these workstations; therefore, no extremely high desktop illuminance was found for this library.

Figure 1. Griffith University Gold Coast Library and its shading strategies: louvers for west facing facades, horizontal and vertical projections for north, south and east facing facades, and trees around the building.

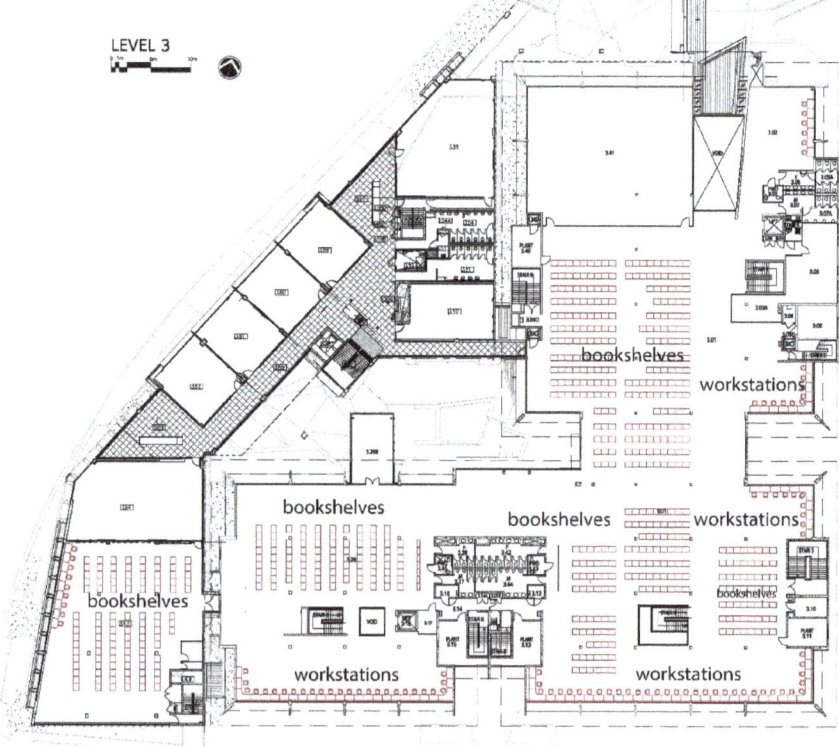

Figure 2. The floor plan of Level 3; 88 workstations which are located next to windows and face the outside are selected for this study.

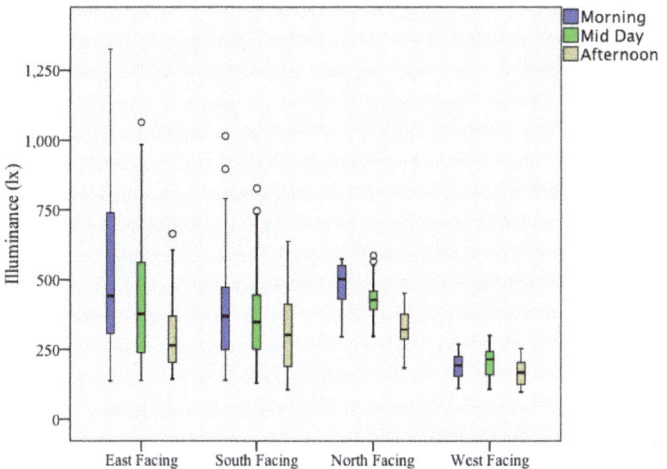

Figure 3. Desktop Illuminance for workstations with different orientations.

2.2. Seat Preference Survey

In response to the first question, a questionnaire survey was conducted to discover the importance of the outdoor view in choosing seats in learning environments. The questionnaire survey was conducted during April–May 2016. The survey is to probe the subjective evaluation of reasons for being seated at some location and the importance of choosing a seat in a learning environment. The questionnaire consisted of three parts: the first part is about the participant's background; the second part is an open question asking the reason that they chose their current seat; and the third part is a structured question ranking factors that contributed to their seat selection. The selected factors include good views, good daylighting, close to toilet/washroom, close to friends/mates, close to reference books, close to entrance/circulation, privacy, quietness, good furniture, clean and orientation. The questionnaire survey was administered to 100 students who were seated in the selected areas. Finally, 72 valid responses were collected. Thirty-nine are female students while 33 are male. Most of them (86.1%) are undergraduate students while few are postgraduate students.

2.3. Quantification of Views and Occupancy Rate

In response to the second question, measurements were conducted to quantify outdoor view elements and to collect occupancy data for the correlation analysis. A Canon EOS 5D Mark III was used to take high dynamic range images for view quantification. A tripod was used to fix the equipment and take photos at each workstation (Figure 4). These photos were taken during holidays in December 2015 to avoid interruptions from library users. The measurement of outdoor views mimicked a person seated at that workstation who was looking at windows. Based on the images taken by the fisheye lens camera, the research conducted view analysis for each workstation using the *Sky View Factor Calculator* developed by Lindberg and Holmer [23]. The calculator can help to quantify the portion of sky, trees, and shading in these hemispherical photographs using a Graphical User interface (GUI). To do so, the photos were processed using Photoshop to highlight the part of trees, sky, and shading, respectively (Figure 5); and then processed photos were imported to the calculator to compute the percentage using a pixel-based approach. The procedure was applied to the view analysis of each workstation. In total, 88 sets of data (percentage of sky view, percentage of greenery view, and percentage of shading view) were collected, representing the main outdoor view conditions for the 88 workstations.

Figure 4. Example of taking high-dynamic-range (HDR) photos at each workstation.

Composition of Outdoor Views   Percentage of Sky View: 12.3%   Percentage of Greenery View: 2.1%   Percentage of Shading View: 10.8%

Figure 5. Example of the view analysis using the Sky View Factor Calculator.

The research used 16 Arduino PIR (Passive Infrared) motion sensors to collect the occupancy data. An electronic circuit was set up to use the PIR motion sensors to collect the occupancy data. The PIR motion sensors can sense a slight motion of the human body and send a signal and trigger occupancy in the place. The data from the sensors were saved in a binary code format: '1' represents detecting at least one occupant in one of the surrounding desks, and '0' represents no occupant in the desks. The delay between triggers was activated at 60 s in order to acquire sufficient occupancy data especially when students stay still while studying. The circuit was placed in a black box and was installed underneath student working desks. PIR's distance sensing range was adjusted to around 5 m to cover 5 to 6 desks. Sixteen locations, which covered all the 88 workstations, were selected to install the circuit (Figure 6).

The data were collected on two days: Monday the 4th and Tuesday the 5th of April 2016. These two days are normal weekdays in the middle semester. Although the PIR motion sensors can continuously record occupy data, the acquired data just indicated general occupancy conditions in the 16 learning spaces. It could not tell the difference of the occupancy condition for each workstation. The data for each workstation should also be recorded manually. Therefore, a research assistant helped to count heads during the first day, Monday 4th April 2016. The head counting was conducted every half hour from 8 a.m. to 8 p.m. The head counting helps to calculate how many occupants were present during daytime and verify the PIR sensors.

2.4. Analysis

The analysis of the data must follow three steps: the first step is to reduce the factors of seat selection through factor analysis, in order to identify the potential variables that could account for students' seat selection in the library; the second step is to investigate the different types of view elements in terms of quantity; and the third step is to link the views and occupancy data acquired in the monitoring via regression analysis, in order to explore the relationships between views and occupancy.

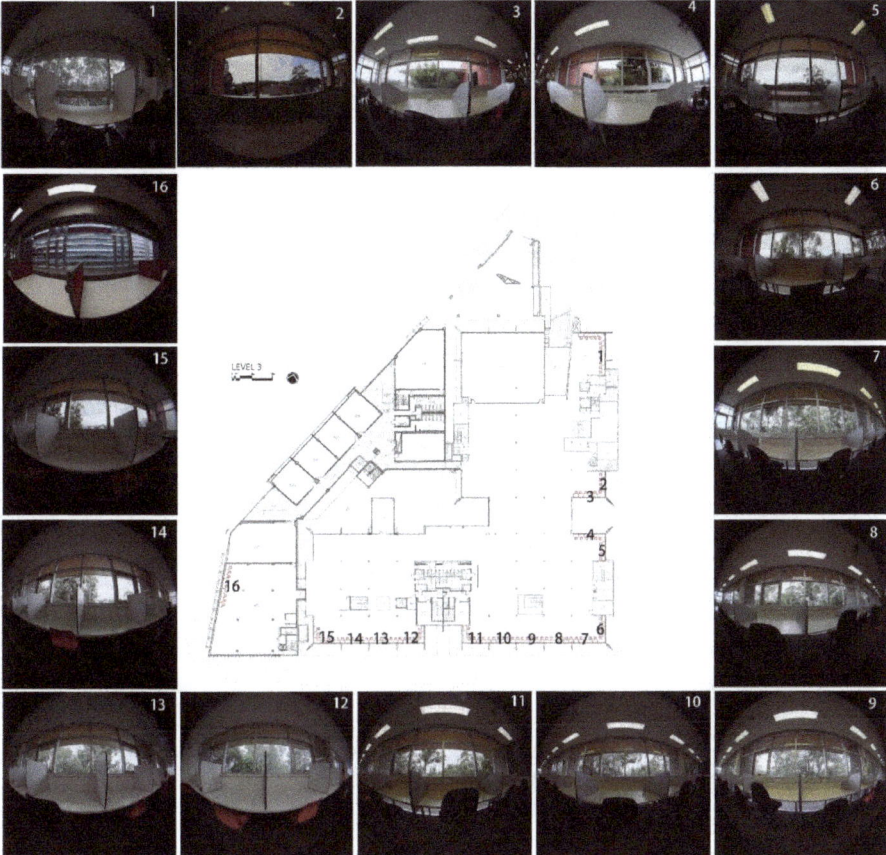

Figure 6. The 16 locations where the Passive Infrared (PIR) motion sensors were installed (the pictures show the outside views at those locations).

3. Results

3.1. Factors Influencing Seat Preference

Figure 7 shows the frequency of reasons mentioned by participants for sitting at their current seats. "Quiet" is most frequently mentioned by participants. "Views" comes in second. The other frequently mentioned reasons are "privacy", "less distraction", and "seclusion". The responses disclose that students prefer some quiet places for concentrating on their learning activities in the library. As assumed, outdoor views are one of the important reasons for choosing their seats. Participants were required to rate the importance of listed factors for choosing seats in the library. Figure 8 disclosed the mean score of these potential factors. As expected, "quiet" was rated as the most important factor. "Furniture" was the second important factor. Convenience, such as "close to friends", "close to reference/books", "close to entrance/circulation", was least important.

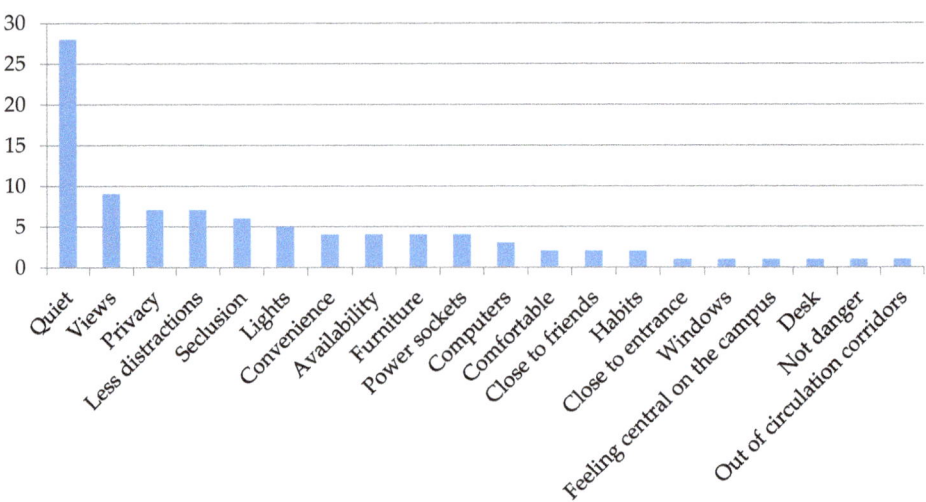

Figure 7. Frequency of reasons being mentioned.

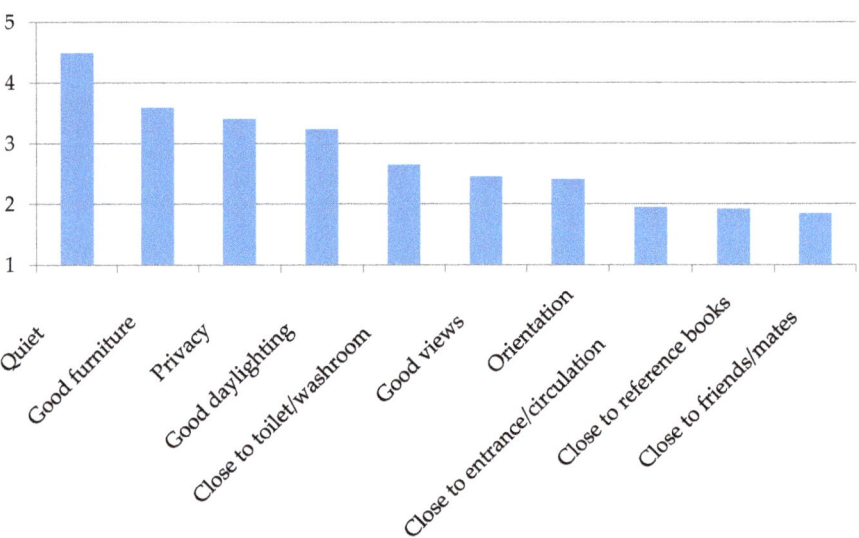

Figure 8. Importance of factors for choosing seats in library.

Factor Analysis was conducted to reduce these factors into several main factors. The Factor Analysis is an explorative analysis, aiming to group similar variables into dimensions. This process is also called identifying latent variables. Three latent variables were extracted for this case, and the three latent variables or main factors could explain 61.2% of variability. Table 1 summarizes the three main factors. Factor 1 includes furniture, privacy, and quietness, which is the most important factor for choosing seats in library; Factor 2 covers views, daylighting and orientation, which is the second most important factor; while Factor 3 refers to being close to friends and entrance, which is the least important among the three main factors.

Table 1. Three underlying factors for choosing seats.

	Factor 1	Factor2	Factor3
	Factor Loading *	**Factor Loading ***	**Factor Loading ***
Furniture	0.680		
Privacy	0.702		
Quiet	0.736		
Views		0.809	
Daylighting		0.685	
Orientation		0.666	
Friends			0.709
Entrance			0.664

* Factor loading shows the relationship of each variable to the underlying factor.

3.2. Views Elements and Their Relationships

Figure 9 shows the percentage of sky, green, and shading view elements for each selected workstation. Since the library building is enclosed by continuous curtain walls, each workstation had a similar size of outdoor views (quantity of outdoor views). However, the composition of outdoor view elements (quality of outdoor views) is quite different. Averagely, these workstations had more shading views (Mean: 12.6%) and green views (12.3%) than sky views (8.4%). The stations with higher percentage sky views are those with less or without trees nearby. Another finding is that the sky and shading views both are negatively associated with the green view (Table 2), which means: the more sky or shading view, the less green view. There is no correlation between sky view and shading view. Figure 10 shows the three extreme conditions: the workstation with the most greenery view, the workstation with the most sky view, and the workstation with the most shading view. In all three conditions, nearby trees and sky are the main outdoor natural elements and they are also negatively associated with each other. The horizontal and vertical shading reduced the sky and green view availability. This is worse when the louvres are present as shading devices, which reduced the natural view size and interrupted its continuity.

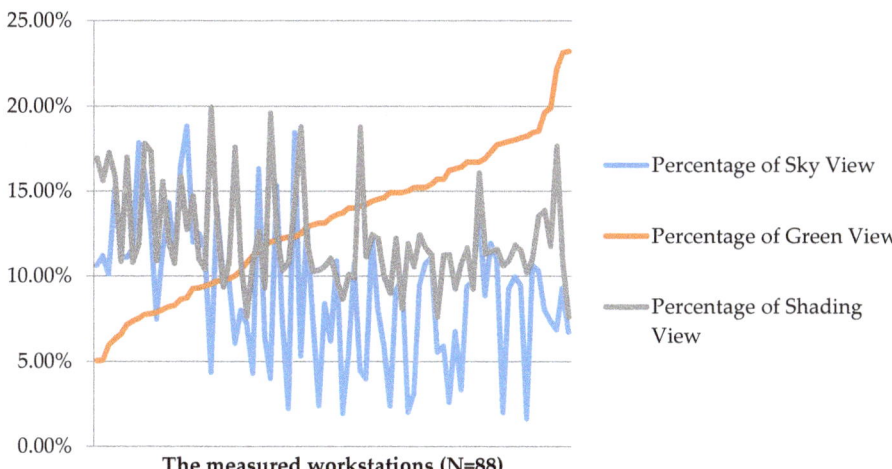

Figure 9. The percentage of sky, green and shading view elements for selected workstations.

Table 2. Correlation table for the three outdoor view elements.

		Percentage of Sky View	Percentage of Green View	Percentage of Shading View
Percentage of Sky View	Pearson Correlation	1	−0.384 **	0.169
	Sig. (2-tailed)		0.000	0.135
	N	88	88	88
Percentage of Green View	Pearson Correlation	−0.384 **	1	−0.325 **
	Sig. (2-tailed)	0.000		0.003
	N	88	88	88
Percentage of Shading View	Pearson Correlation	0.169	−0.325 **	1
	Sig. (2-tailed)	0.135	0.003	
	N	88	88	88

** Correlation is significant at the 0.01 level (2-tailed).

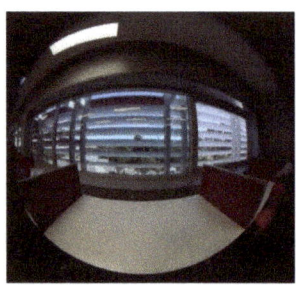 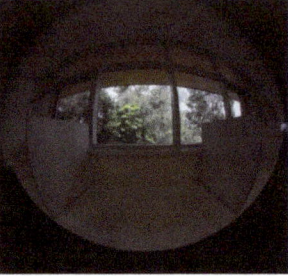 

Figure 10. The workstation with most green view (left) located at 12 in Figure 5, the workstation with most sky view (middle) located at 2 in Figure 5 and the workstation with most shading view (right) located at 16 in Figure 5.

3.3. Linking View Elements to Occupancy Rate

Figure 11 shows the occupancy data for the 16 selected locations representing 16 learning spaces in this library. Two-day data were analyzed. The time period was from 8 o'clock in the morning to 8 o'clock in the evening. These workstations were highly used during the first two days. There is no general pattern that the occupancy rate is greater in the sunny day than in the cloudy day ($p > 0.05$). However, it is found that workstations with south facing had a greater occupancy rate in the sunny day while those with east facing had a greater occupancy rate in the cloudy day. Spaces such as 2, 5, 6, 14, and 15 had higher occupancy rates than others in both two days while spaces 1 and 16 had the least rates. The difference could be initially postulated through comparing outdoor views. Workstations in space 2, 5, 6, 14, and 15 had more sky views while workstations in space 1 and 16 had the least sky views. Usually, there are a number of reasons to choose where to sit, such as noise, privacy, facilities and accessibility. Because the whole floor is a quiet zone and each workstation in this floor was equipped with similar facilities (plug and power), the influence from these factors was supposed to be minimal. The possible reason might be that compared to other locations, space 1 and 16 were to some extent isolated because students need to pass a door to enter the two spaces while other spaces are next to bookshelves.

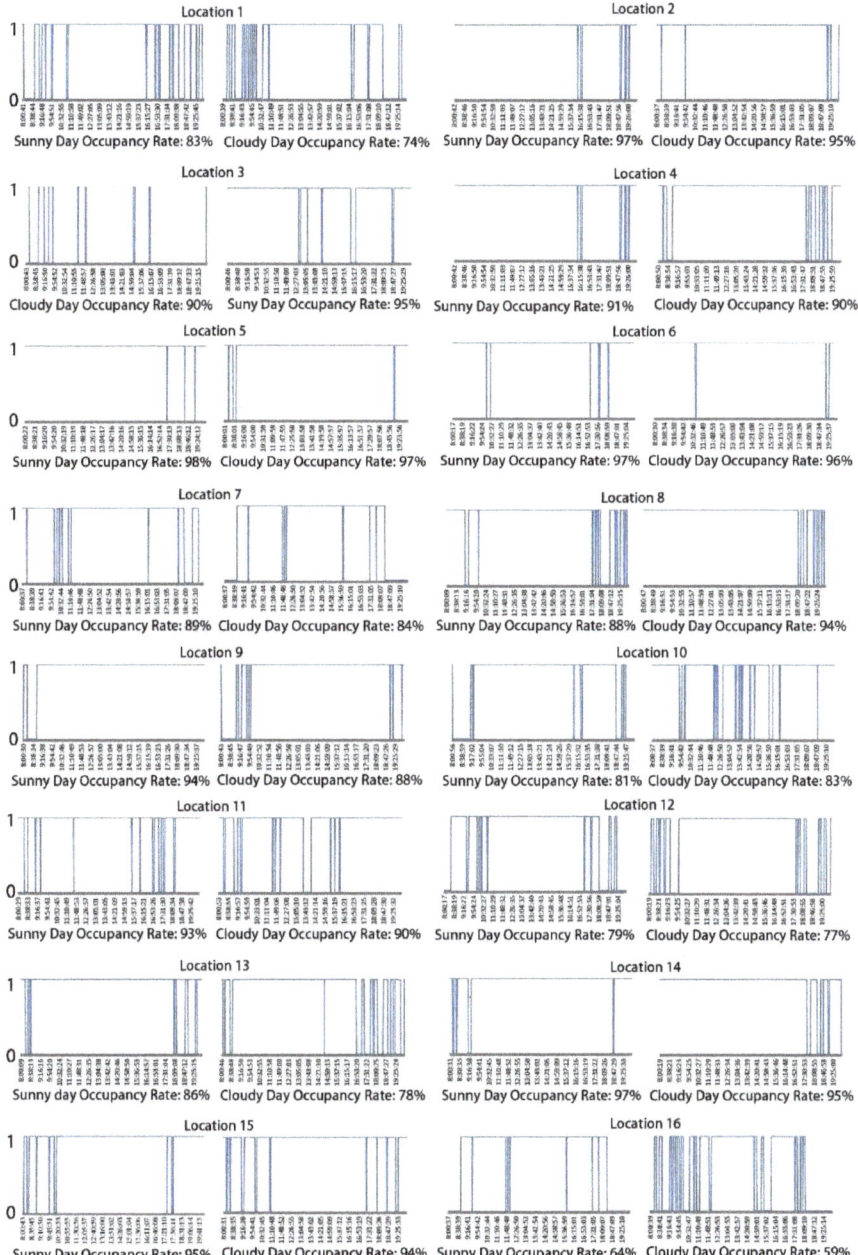

Figure 11. Occupancy rate at each location by PIR sensors (1: occupied; 0: unoccupied).

Figure 12 further breaks down the occupancy data into different time slots. Step-by-step pair tests were conducted to see whether the difference was significant. In both days, the occupancy rate in the midday (11 a.m.–2 p.m.) or afternoon (2–5 p.m.) was significantly higher than that in the morning (8–11 a.m.) or evening (5–8 p.m.) ($p < 0.05$). The peak time in this library was 11 a.m. to 5 p.m.

The breakdown analysis also disclosed that for the least occupied learning spaces such as space 1 and 16, the peak hour occupancy rate was almost the same as the others; the difference mainly came from the non-peak hours. During morning and evening, few students choose to study there. This could exclude the assumption that the two spaces were too isolated to be accessed by students. During peak time, the two spaces were still least occupied by students.

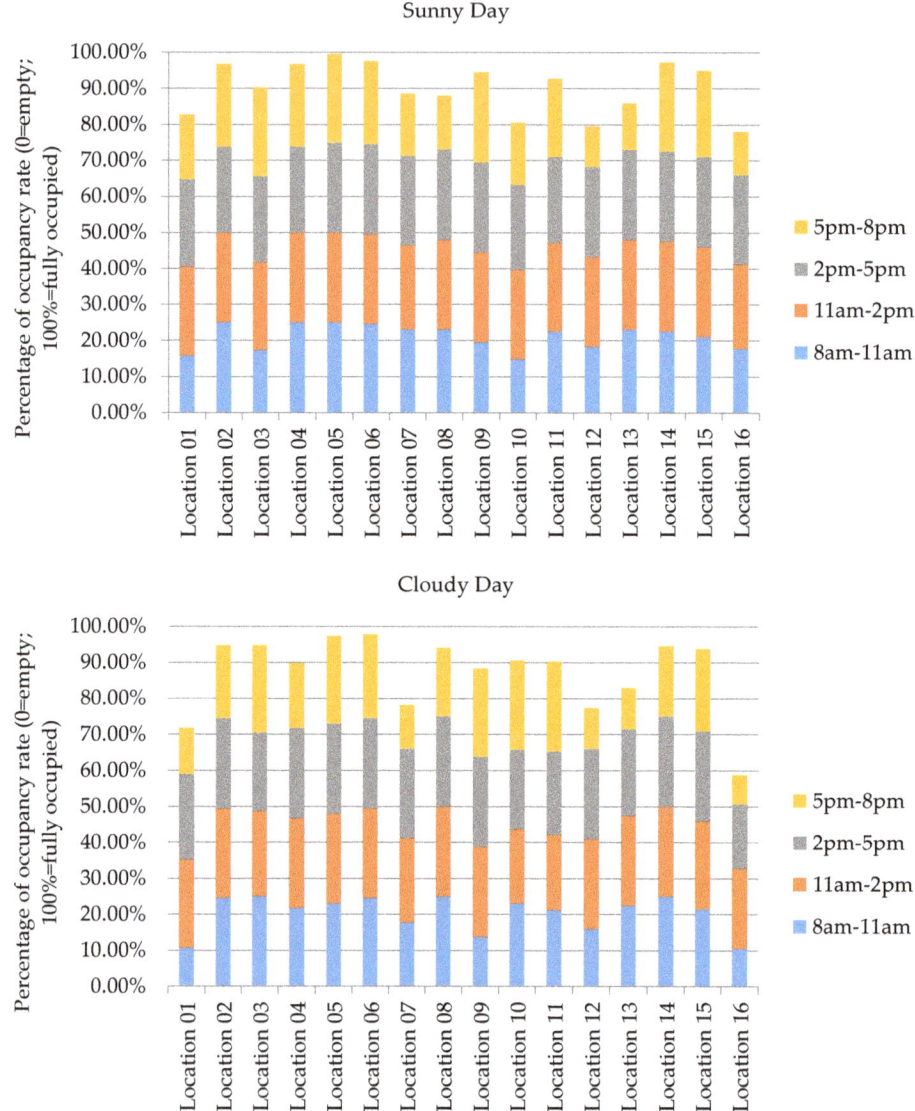

Figure 12. Breakdown of occupancy rates acquired by PIR sensors.

The occupancy data acquired from PIR sensors might not be accurate since each sensor would count nearby occupants. It is necessary to observe the occupancy condition to verify the data. Figure 13 shows the data acquired from one-day observation (heads counting every half hour from 8 a.m.–8 p.m).

The figure also indicates the detection range of each PIR sensor in relation to each workstation. The data disclosed more details of the occupancy rate. The peak occupancy rate was similar to the data acquired by PIR sensors; on the other hand, the valley bottom value was lower than the data acquired by PIR sensors. The PIR seemed to overestimate the occupancy rate due to its detection range. However, the average value could largely match with the data acquired from the PIR sensors.

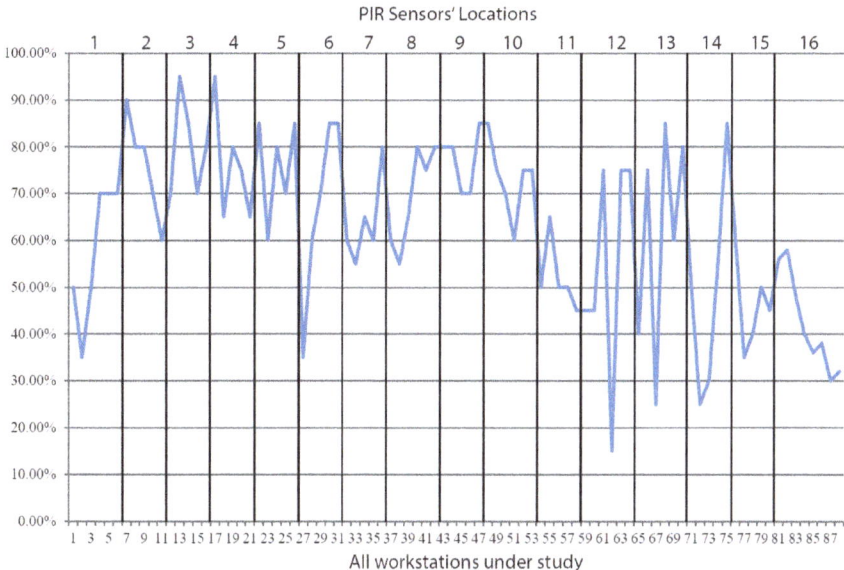

Figure 13. Occupancy rate at each workstation by head counting.

Regression analyses were conducted to explore the relationship between views and users' stay. The dependent variables are the two types of occupancy rate data respectively acquired by PIR sensors and head counting while the independent variables include the average percentage of sky views, average percentage of green view and average percentage of shading views. Table 3 summarizes the two regression models. The two models disclose very similar relationships: sky and shading views tended to positively relate to the occupancy rate while a greenery view was likely to be negatively associated with the occupancy rate. The two regression models support each other to indicate there might be an interesting relationship between views and seat preferences. The relationship deserves further verifications and explorations. The main aim of this regression analysis is to compare the relationships between outdoor view elements (sky, greenery, and shading) and occupancy rate. Therefore, the models are not to claim the individual weighting or contribution of these elements, but to compare their correlations to the occupancy rate.

Table 3. Regression models for occupancy rates.

Dependent Variables	Independent Variables & Coefficients		
	Sky	Greenery	Shading
Occupancy rate_PIR sensors ($R^2 = 0.375$)	0.323 *	−0.197 *	0.287 *
Occupancy rate_Head counting ($R^2 = 0.217$)	0.206 *	−0.116 *	0.178 *

* Correlation is significant at the 0.05 level (2-tailed).

4. Discussion

High quality outdoor views are of great importance not only for improving occupants' health and well-being, but also for attracting and retaining users, which should be addressed in designing learning environments. The seat preference survey disclosed that views are the second most frequently mentioned reason for choosing seats. The factor reduction analysis identified three latent variables accounting for choosing seats: the first is about territoriality (such as quietness, privacy, and furniture) that refers to the personal control of his or her individual space in the library; the second is about visual aspects (such as views, daylighting and orientation); and the third is about social interactions (such as friends, entrance and circulation). Territoriality, in this study, is rated as the most important factor for choosing seats, which is consistent with other research [9]. This study also found that outdoor views came to the second most important factor in students' decision making for seat selection. The outdoor views and related daylight and orientation, are important for occupant health and wellbeing. This study points out that they are important for seat preference as well.

Furthermore, the research investigated three common outdoor view elements: sky, greenery, and shading. These three elements have different effects in a visual environment. Sky and greenery are natural elements that are supposed to have positive effects on attracting users to stay but have contradictive effects on daylight availability; shading is a man-made element that is supposed to be less attractive but necessary for preventing glare or reducing visual contrast. Occupancy rates were found positively associated with sky and shading while negatively related to greenery. This result could be tentatively interpreted that students might prefer sky views with appropriate shading rather than dense trees as outdoor views. The openness to sky can provide more layers of views, which is more interesting than singularity of densely planted trees. The quantification of view elements have not been found in previous literature. This study provided a new approach of using fisheye lens images to quantify different view elements. Three main elements were selected for the quantification. It should be noticed that other elements, such as roofs and grounds, might be a component of outdoor views, and should be included in future studies.

The occupancy rate is an important research subject in this study. Most previous research on daylighting and outdoor view preference used subjective rating, such as perception, and satisfaction as evidence [24]. This research used real occupancy data as evidence to account for users' preference or attraction to stay. Through automatically monitoring and manually counting occupancy rates, a stable difference was found between measured locations and workstations. Workstations with the largest sky views tended to have a higher occupancy rate. The regression model confirmed the three significant factors that influenced the occupancy rate: sky views, tree views, and shading views. It is observable through the occupancy pattern that maximum sky views plus appropriate shading had positive effects on attracting a students' stay. Some studies suggested that layers of views are also important factors contributing to the quality of outdoor views and that views with different layers were preferred over single layer views and a distant city or landscape is the most preferred layer [17]. This point can help to explain that students in this library preferred sky views due to the possibility of including more content and layers while they were less attracted to greenery views because of the singularity.

5. Conclusions

There are many personal and built environment factors influencing users' stay or seat preference in learning environments. This study not only echoes the previous research indicating that territory and privacy are important factors for choosing seats in a learning environment; but also, this study contributes to the literature with evidence that outdoor views might be another important factor for seat preference. Using a library building as a case study, the research investigated three important outdoor view components: sky, greenery, and shading, which may influence daylighting performance and quality views. In this library building, nearby trees are used as an important shading strategy and outdoor view element. However, the high density of leafs and trunks would reduce sky and daylight availability; furthermore, the singularity of view layers would less likely be attractive to

Buildings **2018**, *8*, 96

building occupants. Sky views and shading views were found positively related to the occupancy rate. Based on this point, open views with appropriate shading were found as an optimal outdoor view composition.

The research has important design implications to green building guidelines and practices. Many green building rating systems proposed bringing outdoor views inside for high quality indoor environments; while they did not suggest which view elements should be introduced and in which way different view elements should be composed. The research proposes that an open view with appropriate shading should be attractive to users. The singularity of a view that is less likely to be attractive to building occupants is not recommended.

The outdoor view is usually entangled with daylight; that is why many green building standards combine daylight and outdoor views as one of the important indoor environment quality design aspects. Therefore, it is recommended that future studies shall look at the interactive effects of daylight and outdoor views to enrich the existing daylighting design metrics [25,26].

Methodologically, this study provides new approaches to quantify view elements and to count the occupancy rate using motion sensors. However, the study contains several limitations. The first is that the seat selection was focused on the periphery of the library space while excluding the middle seats in the library. Although it is intentional since the periphery seats have direct outdoor views to study the effect, it is more interesting to see whether the effect applies to seats in other locations in the library. The second is about the sample size for the survey. Principally, 100 respondents are expected to conduct the factor analysis. However, only 72 students returned the completed survey results. The third is about the occupancy monitoring. Only two weekdays were monitored. Due to the limitations, the results and findings should be carefully generalized. Future studies are expected to include a larger sample size and longer days to monitor the occupancy conditions.

Author Contributions: Conceptualization, Methodology, Formal Analysis, Supervision and Draft Writing by Z.G.; Investigation, Data Curation and Writing Review by M.K. and B.M.

Funding: This research received no external funding.

Acknowledgments: The authors would like to thank the Griffith University library for the tremendous help in the data collection.

Conflicts of Interest: The authors declared no potential conflicts of interest with respect to the research, authorship, and/or publication of this article.

References

1. Weinstein, C.S. The physical environment of the school: A review of the research. *Rev. Educ. Res.* **1979**, *49*, 577–610. [CrossRef]
2. Bluyssen, P.M. Health, comfort and performance of children in classrooms—New directions for research. *Indoor Built Environ.* **2017**, *26*, 1040–1050. [CrossRef]
3. Haghighi, M.M.; Jusan, M.B.M. The impact of classroom settings on students' seat-selection and academic performance. *Indoor Built Environ.* **2013**, *24*, 280–288. [CrossRef]
4. Yildirim, K.; Cagatay, K.; Ayalp, N. Effect of wall colour on the perception of classrooms. *Indoor Built Environ.* **2014**, *24*, 607–616. [CrossRef]
5. De Giuli, V.; Zecchin, R.; Corain, L.; Salmaso, L. Measurements of indoor environmental conditions in Italian classrooms and their impact on children's comfort. *Indoor Built Environ.* **2014**, *24*, 689–712. [CrossRef]
6. Mahyuddin, N.; Awbi, H.B.; Alshitawi, M. The spatial distribution of carbon dioxide in rooms with particular application to classrooms. *Indoor Built Environ.* **2013**, *23*, 433–448. [CrossRef]
7. Wang, N.; Boubekri, M. Investigation of declared seating preference and measured cognitive performance in a sunlit room. *J. Environ. Psychol.* **2010**, *30*, 226–238. [CrossRef]
8. Galasiu, A.D.; Veitch, J.A. Occupant preferences and satisfaction with the luminous environment and control systems in daylit offices: A literature review. *Energy Build.* **2006**, *38*, 728–742. [CrossRef]

9. Kaya, N.; Burgess, B. Territoriality: Seat preferences in different types of classroom arrangements. *Environ. Behav.* **2007**, *39*, 859–876. [CrossRef]

10. Gifford, R. *Environmental Psychology: Principles & Practice*; Optimal Books: Colville, WA, USA, 2002.

11. Pedersen, D.M. Privacy preferences and classroom seat selection. *Soc. Behav. Personal. Int. J.* **1994**, *22*, 393–398. [CrossRef]

12. Ulrich, R. View through a window may influence recovery from surgery. *Science* **1984**, *224*, 420–421. [CrossRef] [PubMed]

13. USGBC. *LEED v4 for Building Design and Construction*; U.S. Green Building Council: Washington, DC, USA, 2014.

14. Boubekri, M. *Daylighting, Architecture and Health: Building Design Strategies*; Architectural Press: Oxford, UK, 2008.

15. Gou, Z.; Lau, S.S.-Y.; Qian, F. Comparison of mood and task performance in naturally-lit and artificially-lit environments. *Indoor Built Environ.* **2015**, *24*, 27–36. [CrossRef]

16. Andersen, M. Unweaving the human response in daylighting design. *Build. Environ.* **2015**, *91*, 101–117. [CrossRef]

17. Hellinga, H.; Hordijk, T. The D&V analysis method: A method for the analysis of daylight access and view quality. *Build. Environ.* **2014**, *79*, 101–114.

18. Edwards, L.; Torcellini, P. *A Literature Review of the Effects of Natural Light on Building Occupants*; National Renewable Energy Laboratory: Golden, CO, USA, 2002.

19. Lau, S.S.Y.; Gou, Z.; Liu, Y. Healthy campus by open space design: Approaches and guidelines. *Front. Archit. Res.* **2014**, *3*, 452–467. [CrossRef]

20. Velarde, M.D.; Fry, G.; Tveit, M. Health effects of viewing landscapes—Landscape types in environmental psychology. *Urban For. Urban Green.* **2007**, *6*, 199–212. [CrossRef]

21. Xue, F.; Gou, Z.; Lau, S. Human factors in green office building design: The impact of workplace green features on health perceptions in high-rise high-density Asian cities. *Sustainability* **2016**, *8*, 1095. [CrossRef]

22. Othman, A.R.; Mazli, M.A.M. Influences of Daylighting towards Readers' Satisfaction at Raja Tun Uda Public Library, Shah Alam. *Procedia Soc. Behav. Sci.* **2012**, *68*, 244–257. [CrossRef]

23. Lindberg, F.; Holmer, B. *Sky View Factor Calculator: User Manual—Version 1.1*; Göteborg Urban Climate Group, University of Gothenburg: Gothenburg, Sweden, 2012.

24. Reinhart, C.F.; Mardaljevic, J.; Rogers, Z. Dynamic daylight performance metrics for sustainable building design. *Leukos* **2006**, *3*, 7–31.

25. Nabil, A.; Mardaljevic, J. Useful daylight illuminance: A new paradigm for assessing daylight in buildings. *Light. Res. Technol.* **2005**, *37*, 41–57. [CrossRef]

26. Salata, F.; Golasi, I.; di Salvatore, M.; de Lieto Vollaro, A. Energy and reliability optimization of a system that combines daylighting and artificial sources. A case study carried out in academic buildings. *Appl. Energy* **2016**, *169*, 250–266. [CrossRef]

 © 2018 by the authors. Licensee MDPI, Basel, Switzerland. This article is an open access article distributed under the terms and conditions of the Creative Commons Attribution (CC BY) license (http://creativecommons.org/licenses/by/4.0/).

Article

Thermal Comfort Analyses of Secondary School Students in the Tropics

Baharuddin Hamzah [1,*], Zhonghua Gou [2], Rosady Mulyadi [1] and Samsuddin Amin [1]

[1] Department of Architecture, Hasanuddin University, Makassar 90245, Indonesia; rosady@unhas.ac.id (R.M.); masarchiuh@yahoo.com (S.A.)
[2] School of Engineering and Built Environment, Griffith University, Gold Coast, QLD 4215, Australia; z.gou@griffith.edu.au
* Correspondence: baharsyah@unhas.ac.id; Tel.: +62-852-5563-7065

Received: 1 March 2018; Accepted: 7 April 2018; Published: 10 April 2018

Abstract: This study aims to analyze the thermal comfort level of students in secondary schools in the tropical city of Makassar. The analysis is carried out based on data surveyed from eight selected high schools. The study involved 1594 students in 48 classrooms. The recorded data includes personal data and measured environmental parameters. At the same time, students were asked to fill out questionnaires related to their thermal comfort levels. The surveyed classrooms showed high air temperatures. The air temperatures ranged from 28.2 °C in the morning to 33.6 °C in the midday. The radiant temperatures were similar to the air temperature, which indicated that the airflow speed was low. The only parameter that could meet the Indonesian national standard was relative humidity. However, many students still feel comfortable (−1 to +1) based on TSV (thermal sensation vote) and TCV (thermal comfort vote). Even though about 80% of respondents accepted this hot temperature, most of them preferred to have a decrease in the air temperature. Regarding the PMV (predicted mean vote), only about 23% respondents were predicted to feel slightly warm (+1). The regression analyses show that the neutral temperatures were 29.0 °C and 28.5 °C for TSV and TCV, respectively.

Keywords: thermal comfort; natural ventilation; measurement; school classrooms; secondary school

---

1. Introduction

Thermal comfort is defined in the American Society of Heating, Refrigerating and Air-Conditioning Engineers (ASHRAE) 55 standard as the "that condition of mind which expresses satisfaction with the thermal environment" [1]. This definition is later adopted by International Standard Organization (ISO) in its Standard 7730 [2]. According to Gagge et al. [3] the sense of comfort is complex because it applies the entire environment including all the psychological and physiological aspects. In details, Fanger [4] explained that thermal comfort is determined by several factors, including thermal environments, personal factors, and other contributing factors. Environmental factors include air temperature, air velocity, humidity, and radiation. Personal factors include the clothing and the activity (metabolic rate). Contributing factors include food and drink, acclimatization, body shape, subcutaneous fat, age and sex, and state of health. During his live, late Professor P.O. Fanger has extensively carried out research involving most of the variables mentioned above. The summary of his works can be read in the recent work done by d'Ambrosio Alfano et al. [5]. However, there are still a lot of unsolved problems in the thermal comfort studies, especially in the tropic area, such as in Indonesia.

Thermal comfort is one of the important environmental factors for building occupants to work productively and live well [6,7]. Several literatures found that there was a positive correlation between thermal environments and productivity of workers in office buildings [8,9]. Occupants' well-being and comfort are dependent on the indoor environmental quality [10]. Similar to the office buildings, thermal comfort is also becoming an important requirement in educational buildings for enabling students learn to productively. A quite old study in 1968 revealed that there was a positive effect of the thermal quality of classrooms on students' performance [11]. According to d'Ambrosio Alfano et al. [12] providing a comfortable and healthy environment in school is necessary for students well-being and productivity. Pepler and Warner [11] reported the result from an experimental study of learning efficiency among adults in a school laboratory as the temperatures increased from 17 to 27 °C. An extensive literature review by Mendell and Heath [13] showed a good correlation between indoor school settings and the performance and attendance of students. They also found that warmer temperatures (above 24 °C) tended to reduce the performance of students.

Numerous researchers have conducted studies on thermal comfort of students in the classrooms in various places in the world. These studies include the analysis of thermal comfort in schools in the temperate climate in the United Kingdom (UK) [14–16], Mediterranean climate in Italy [17–21], subtropical climate in Taiwan [22–25], Japan [26,27], and Australia [28,29]. In the tropical area, studies of thermal comfort at schools have been made in Singapore [30], Malaysia [26] and Hawaii [31].

A result of the study was carried out in the United Kingdom (UK) by Teli et al. [14] suggests that children are more sensitive to higher temperatures than adults with the comfort temperatures being about 4 °C and 2 °C lower than the PMV (predicted mean vote) and the EN 15251 adaptive comfort model predictions, respectively. In Italian naturally ventilated schools, d'Ambrosio Alfano et al. [21] found a good agreement between PMV and TSV (thermal sensation vote) with expectation factor e 0.9. They found that the use of Fanger's basic approach in calculating thermal comfort could also be applied in naturally ventilated environments if the right expectancy factor is known. Hwang et al. [23] carried out an extensive field measurement in air conditioning (AC) and naturally ventilated (NV) classrooms in Taiwan. The survey involved 944 students at 36 classrooms (ten of them were naturally ventilated classrooms). They found that the thermal neutrality and thermal preference of students were 26.3 and 24.7 °C T_e, respectively. This result shows that the neutral temperature was higher than the thermal preference of students. Also, the study shows that the neutral temperature in Taiwan was lower than the neutral temperature found in the secondary school in Singapore [30]. The PMV model overestimated the TSV of students. Kwok and Chun [27] carried out a research in high schools in Japan to determine students' thermal comfort. They found that the thermal environments in the surveyed schools were beyond the thermal comfort zone specified in the ASHRAE standard. The average air temperature was 26.9 °C and MRT was 27.1 °C. However, about 72% of respondents found this condition acceptable. Most of the students (74%) voted within the neutral category ("slightly cool", "neutral", and "slightly warm"), while only less than 10% voted for "warm" and "hot". Based on a large data gathered from the survey at six elementary and three high schools in Australia, de Dear et al. [28] found the neutral and preferred operative temperatures were about 22.5 °C, which falls below predictions of both PMV and adaptive models of thermal comfort. The temperatures at that time were 18.2–31.1 °C, with an average value of 25.1 °C. They also found that children's thermal sensation and temperature preference drop 1–2 °C below those of adults. Kim and de Dear [29] study the applicability of adaptive thermal comfort model in the Australian primary and secondary school students. They found that more than 80% of students in primary (89.2%) and secondary (86.0%) school accept the indoor operative temperature of 24.5 °C and 24.7 °C in primary and secondary, respectively.

Based on a survey carried out in the Singaporean secondary schools, Wong and Khoo [30] found that none of the thermal performances of classrooms were within the thermal zone of the ASHRAE standard. However, students found these conditions acceptable. The acceptability rates were 72% and 74% for NV and AC classrooms, respectively. The neutral temperature found in this study was 28.8 °C. The neutral temperature predicted by the PMV model was higher than the one obtained from actual votes TSV. Kwok [31] examined the acceptability of the ASHRAE thermal comfort standard for the tropical classroom in Hawaii. Kwok found that the majority of classrooms failed to meet the ASHRAE standard. However, the acceptability rate was more than 80% irrespective of the thermal condition of classrooms. The neutral temperature values in these tropical classrooms were 26.8 and 27.1 °C for NV and AC classrooms, respectively.

Unfortunately, research on the thermal comfort of students in the secondary schools still lacks in Indonesia, especially in Makassar. Makassar, the provincial capital of South Sulawesi, is the largest city on Sulawesi Island in terms of population, and the fifth largest city in Indonesia. The city has a tropical monsoon climate with the average temperature for the year at 27.5 °C, the average high around 32.5 °C and the average low around 22.5 °C. There are few thermal comfort studies in tropical Indonesia. A thermal comfort study by Feriadi and Wong [32] is focused on residential buildings in Jogyakarta. Feriadi and Wong [32] showed that the prediction of thermal comfort using a PMV model overestimated the thermal sensation vote (TSV) and the thermal comfort vote (TCV) of the respondents. More than 95% of the respondents were predicted by the PMV method to have thermal sensation in the warmer region(+1 to +3) and only very little (less than 5%) in the "neutral" to cooler region (0 to −3). Karyono [33] carried out thermal comfort study in air conditioned office buildings in Jakarta. Karyono's study was based on an extensive survey of 596 office workers from seven high-rise office buildings in Jakarta. The study showed that most of the office workers were still comfortable in room temperatures between 26.7 and 28.6 °C. Karyono found a neutral temperature of 26.7 °C (T_o) for air conditioned (AC) office buildings in Jakarta.

Recent studies in Indonesia carried out by Hamzah et al. [34] and Karyono et al. [35] were focused on the university classrooms. Hamzah et al. [34] investigated the naturally ventilated classrooms, while Karyono et al. [35] studied the air conditioning classrooms. Hamzah et al. [34] found that the thermal condition of the classrooms did not meet the requirement of ASHRAE and SNI standard. The major finding of this study is that more than 80% voted the central position (−1 to +1), either in ASHRAE or Bedford scale and the neutral temperature about 29.6 °C. Karyono et al. [35] found that comfort temperature was 24.1 °C T_a and 24.9 °C T_a for students at Universitas Tarumanegara (Untar) and Universitas Mercu Buana (UMB), respectively. Both figures are very low in comparison to the naturally ventilated classroom in the study carried out by Hamzah et al. [34].

Thermal comfort standards such as ASHRAE Standard 55 [36] has been widely used as a guideline for designing thermal comfort in different countries. The measurement of the thermal comfort experienced by the users is usually according to the ASHRAE standard, using a questionnaire based on a study conducted by Fanger [4]. This survey asks the sensation of thermal perceived users in seven scales, namely: hot (+3), warm (+2), slightly warm (+1), neutral (0), slightly cool (−1), cool (−2), and cold (−3). Bedford [37] has proposed a method of measuring thermal comfort in buildings. It also consists of seven scales: much too warm (+3); too warm (+2); comfortably warm (+1); comfortable (0); comfortably cool (−1); too cool (−2), and; much too cool (−3).

Most schools in Indonesia were built as a prototype building, with no consideration based on the local climatic conditions. The schools were built to the same standards regardless of the user's comfort and preference, in this case, the students of the secondary school, which is in the phase of changing from childhood to adulthood (11–18 years). There is no special regulation for the thermal comfort in the educational buildings in Indonesia. The government only provides thermal comfort guidelines for buildings in general. The requirement is based on the Mom and Wiesebron [38] which later on used in the national standard SNI 03-6572-2001 [39], where the building should provide the following thermal environment as follows:

1. comfortably cool: 20.5–22.8 °C (T_e);
2. comfortable: 22.8–25.8 °C (T_e); and
3. comfortably warm: 25.8–27.1 °C (T_e).

The guideline used the effective temperature (T_e), which is defined as the temperature of a stagnant and saturated atmosphere, which would, in the absence of radiation, produce the same effect as the atmosphere in an inquiry. Therefore, it combines the effect of dry air temperature and humidity [40].

Most of the classrooms in the secondary schools in Makassar are experiencing hot temperature during daytime. Through observations, we found that a lot of numbers of classrooms are equipped with fans and even air conditioning (AC). In order to improve the situation and considering the importance of thermal comfort in affecting the students' learning process and performances, then a study of thermal comfort in the classrooms of secondary schools need to be undertaken. The objectives of this study are:

1. To report the thermal environmental conditions of classrooms in the secondary schools in Makassar;
2. To analyze the students' responses to the thermal environmental conditions in their classrooms;
3. To find out the neutral and comfortable temperature in the classrooms of secondary schools based on the climate of Makassar.

2. Research Methods

2.1. Research Sample and Respondent

The present study was conducted at the 48 classrooms, which are used for the teaching and learning purpose. The classroom buildings are mostly one to two stories buildings. There are two types of state secondary schools: State Junior High School for year 7 to 9 (*Sekolah Menengah Pertama Negeri*, which is abbreviated as SMPN) and State Senior High School for year 10 to 12 (*Sekolah Menengah Atas Negeri*, which is abbreviated as SMAN). The locations of schools are spread out from the city center with busy streets and high-density settlements to the suburbs with lower density settlements. Table 1 shows the characteristic of surveyed schools and samples. The schools were chosen to represent the six sub-districts in the city of Makassar, by considering the geographical conditions, and density of buildings. Also, the accessibility to each school is one of the main considerations in this selection.

Table 1. The number of classes and students (respondents) of each school.

No.	Schools Name	Location (Sub-District)	Number of Classes	Number of Students	Date of Survey
1	SMPN 33 Makassar	Rappocini	6	210	1 August 2017
2	SMAN 4 Makassar	Ujung Tanah	7	226	2 August 2017
3	SMAN 21 Makassar	Tamalanrea	6	196	3 August 2017
4	SMAN 1 Makassar	Bontoala	6	197	4 August 2017
5	SMPN 20 Makassar	Manggala	6	186	7 August 2017
6	SMPN 30 Makassar	Tamalanrea	6	212	8 August 2017
7	SMPN 8 Makassar	Manggala	6	204	9 August 2017
8	SMAN 2 Makassar	Mamajang	5	163	11 August 2017
	Total		48	1594	

The specific areas and descriptions of each school and classrooms are explained as follows. SMPN 33 Makassar is located in a high-density settlement. The weather condition was mostly sunny. The measurements were conducted from 8:30 to 13:50. Measurements were carried out in the six classes, where two classes were located on the ground floor, while four classes on the first floor. The size of the class is 63 and 56 m^2. The height of the ceilings range from 2.8 m to 3.6 m. The opening in the classroom was located on the left and right sides of the class with the total window area and total door area of 21.7 m^2. The natural ventilation is accessible through jalousies at the top and the openable window underneath.

SMAN 4 Makassar is located on Jalan Cakalang, about 2.5 km north of the city center. The weather condition on that day was mostly sunny. The measurement of the microclimate conditions in the classroom started at 8:30 and finished at 12:25. Measurements were conducted in seven classrooms. Six of classes are located on ground floor and only one class on the first floor. The class size was varied between 68, and 72.35 m². The ceiling heights range from 3 m to 3.9 m. The opening in the class is located on the left and right sides of the class with the average window area on the right side of 11.75 m² and the left side of 7 m². The natural ventilation can be felt through jalousies at the top and the openable window underneath.

SMAN 21 Makassar is found in a periphery of the dense residential area bordering the road environment. The site was surrounded by less dense trees. The survey and measurements were started in the morning at 8:10 a.m. until noon at 1:45 p.m. The weather condition on the day of measurement was sunny. Measurements were made in six classes consisting of four North-South oriented classes and two East-West oriented classes. Three classes are located on ground floor, while other three on the first floor. The area of each class is 72 m² (8 m × 9 m). The buildings are permanent building type with a plastered brick wall, and concrete tiles roof.

SMAN 1 Makassar is established in the area that is very close to the city center and adjacent to arterial roads. The school's site has a lot of trees. The measurements were conducted in the six classes from 7:55 a.m. to 11:30 a.m. Two classes are located on the ground floor, two classes on the first floor, and one class on the second floor. The class size of the survey object varied: 63, 72, 96, and 100 m², respectively. The height of the ceiling of the classrooms ranges from 3 m to 3.5 m. The window opening in the class is located on the left and right sides of the class with the average area of the window on the right side is 10 m² and the left side is 7 m². The natural ventilation is accessible through jalousies at the top and a glass window underneath.

SMPN 20 Makassar is located in the suburban area with less dense residential areas. The site has less vegetation in surrounding buildings. The weather condition was mostly sunny. The measurement of microclimate in the classroom was done from 8:20 a.m. to 12:00 p.m. The classroom area was varied, between 60 and 62.4 m². The ceiling height of the classrooms ranges from 3 m to 3.5 m. The opening in the classroom is located on the left and right sides of the class with the average window area on the right side of 13.4 m² and the left side is 6.26 m². The natural ventilation made use of jalousies at the top and openable window underneath.

SMPN 30 is located in the high-density residential areas and bordering by busy roads. The measurement and survey were conducted in the morning at 8:15 a.m. until afternoon at 2:40 p.m. Measurements were made in six classes consisting of three Southeast-Northwest oriented classes and three Northeast-Southwest oriented classes. Four classes are located on the ground floor and two on the first floor. The size of each classroom is 7 m × 9 m (63 m²). The wall construction was the plaster-brick wall, and the roof using tile, asbestos, and zinc.

SMPN 8 Makassar is located in the medium density residential and commercial areas. In general, the weather conditions on that day was sunny with some cloudy. The measurement of microclimate in the classroom was done from 8:30 a.m. to 1:05 p.m. Measurements were made in six classes, four classes located on the ground floor and two classes located on the first floor. Except for one class, the size of the five classes was 63 m². The ceiling height of the classroom ranges from 3 m up to 3.5 m. The opening in the class was located on the left and right sides of the class with the average area of the window on the right side of 10 m² and the left side is 8.5 m².

SMAN 2 Makassar is also found in the medium density residential and business areas. The weather condition on that day was sunny. The measurements were conducted from 8:30 a.m. to 11:30 a.m. Measurements were carried out in five classes, one class on the ground floor, two classes on the first floor, and one class on the second floor. The size of the classes was varied, 63 m² and 72 m². The height of ceiling was ranging from 3.2 m to 3.5 m. The opening in the classroom is located on the left and right sides of the class with the total window and open door area is 20.9 m².

The distributions of respondents based on their age and sex are presented in Table 2. As seen in the table, a number of female students are bigger than the male one. Basically, the normal age for junior high school is 13 to 15 years and senior high school is 16 to 18. The reason for a little number of students in the 18 years group and a big number in the 11 and 12 group is may be caused by the time of the survey. The survey was conducted in August, where the new academic year was just begun. Also, some students started their primary school at the age of five years and finished it in 11 years old.

Table 2. The distribution of respondents based on their age and sex.

Sex	Age (Year)								Total
	11	**12**	**13**	**14**	**15**	**16**	**17**	**18**	
Male	12	78	128	111	88	113	67	2	599
Female	14	112	183	168	176	214	123	5	995
Total	26	190	311	279	264	327	190	7	1594

2.2. Research Instrumentation

The research has been carried out using several instruments. The LSI-Lastem Thermal Comfort Multi Logger is a set of devices, which consists of several sensors and data loggers. The arrangements of LSI-Lastem applied in this survey including one data loggers and four sensors. The sensors including a globe thermometric probe (EST131) for measuring mean radiant temperature (MRT). A portable psychometric forced ventilation probe (ESU102) for measuring air temperature and relative humidity, and the hot wire anemometer (ESV106) for recording the air velocity in the classroom.

Figure 1 shows the arrangement of instruments inside the surveyed classroom (left), and the images of instruments used in the survey (right). Because there is only one set of LSI-Lastem logger, additional instruments were needed. Six HOBO loggers made by Onset, have been used for this research. These instruments enabled us to measure the thermal environments at six points in each classroom. Two types of loggers were used, that is, the HOBO temp/Relative Humidity (RH) logger (Hobo-1) and the HOBO temp/RH/Light/External logger (Hobo-2). Four HOBO temperature/RH loggers were used for measuring air temperature and relative humidity, and two HOBO temp/RH/Light/External were used for measuring air temperature, relative humidity, and airflow velocity. The specifications of the instruments used in the data collection are displayed in Table 3.

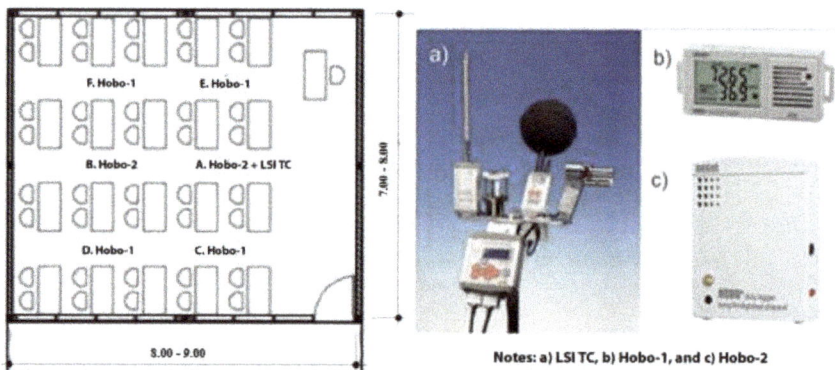

Figure 1. The arrangement of instruments in the typical classrooms (**left**) and the instruments (**right**): (**a**) LSI-Lastem; (**b**) Hobo-1, Hobo temp/Relative Humidity (RH); and (**c**) Hobo-2, Hobo temp/RH/Light/External logger (airflow velocity).

Table 3. The specification of instruments used in the surveys.

No.	Instrument Name	Range	Accuracy	Resolution
1.	Black Globe Radiant temperature (EST131)			
	- Radiant temperature	−40 to +80 °C	NA	0.01 °C
2.	Psychrometer Sensor (ESU102)			
	- Air temperature	−5 to +60 °C	NA	0.01 °C
	- Relative humidity	0 to 100%	NA	1%
3.	Hot wire anemometer (ESV106)			
	- Air velocity	0.01 to 20 m/s	NA	0.01 m/s
4.	HOBO Temp/RH logger (UX100-011)			
	- Air temperature	−20 to +70 °C	±0.21 °C	0.024 °C
	- Relative humidity	5 to 95%	±2.5%	0.05%
5.	HOBO Temp/RH/Light/External (U12-012)			
	- Air temperature	−20 to +70 °C	±0.21 °C	0.024 °C
	- Relative humidity	5 to 95%	±2.5%	0.05%
	- Air velocity	0.15 to 10 m/s	±0.05 m/s	NA

Note: NA = Not available.

2.3. Data Collection

The primary data has been collected through survey and questionnaire methods. The collection of data was carried out as follows:

1. Survey on objective measurement was conducted to collect the personal and the thermal environment data. Personal data was gathered by collecting the clothing and the activity of each respondent. The sensors for measuring the environmental data were attached at 100 cm above the floor level [30,34]. Because of the limited number of equipment, the MRT (Meant Radiant Temperature) was only recorded at one point that was the center of the room, while the air velocity, air temperature, and the relative air humidity were recorded at two points (A and B). The instruments placed in other points C, D, E, and F only measured the air temperature and the relative air humidity (see Figure 1).

2. Survey on subjective measurement was conducted to measure the level of thermal comfort of respondents. The survey carried out by using questionnaire technique, which was adapted from Wong and Khoo and has been used in the previous study [30,34]. The questionnaire included seven questions, which captured the thermal sensation vote (TSV), thermal comfort vote (TCV), thermal preference and thermal acceptance of respondents. In addition to the air temperature, the questionnaire also intended to obtain the respondents' votes on the air velocity, air velocity preference as well as the humidity of classrooms. The TSV responses were measured based on ASHRAE standard 55, which uses a seven-point scale to measure the thermal sensation of respondents. The thermal comfort can also be measured by asking the thermal preference and acceptance of occupants. Thermal preference related to the question of whether the occupants prefer to be warmer or cooler or no change. In addition, questions related to the air velocity and the humidity had also been included in the questionnaire (Table 4). In the top part of the questionnaire, respondents are requested to write down his/her school name, class, student name, sex, age, clothing ensembles, weather condition, and his/her position in the classroom. Students fill out the questionnaire after at least 25 to 30 min sitting in the classroom. In order to prevent the error in choosing the relevant answer based on their feelings and preferences, an explanation on the indicator used in the questionnaire has been carried out, for example, the difference between "cold", "cool", "neutral", "warm", and "hot".

Table 4. Thermal comfort questionnaire (adapted from [30,34]).

1. How do you feel about the temperature in the classroom at this moment?
□ cold
□ cool
□ slightly cool
□ neutral
□ slightly warm
□ warm
□ hot
2. Do you comfortable now?
□ much too cool
□ too cool
□ comfortably cool
□ comfortable
□ comfortably warm
□ too warm
□ much too warm
3. What do you like to be?
□ cooler
□ no change
□ warmer
4. How do you rate the overall acceptability of the temperature at this moment?
□ acceptable
□ not acceptable
5. How do you feel about the air velocity in the classroom at this moment?
□ too still
□ slightly still
□ just right
□ slightly breezy
□ too breezy
6. What do you like to be about the air velocity?
□ increase velocity
□ no change
□ decrease velocity
7. How do you feel about the humidity in the classroom at this moment?
□ much too humid
□ too humid
□ slightly humid
□ just right
□ slightly dry
□ too dry
□ much too dry

The situation of survey and measurement in the selected classrooms is shown in Figure 2. The figure shows the students clothing and activities during the survey and measurement.

Typically, secondary students have five types of uniform that is, regular uniform, *batik* uniform, *pramuka* uniform, Moslem uniform, and the sports uniform. The regular, *batik* and *pramuka* uniforms basically have the same clothing ensembles. They are only different in terms of clothing's color. The regular uniform for male students is a light short-sleeves shirt with light trousers, which has a clothing insulation of 0.57 clo. While the regular uniform for female students is light long-skirt (ankle-length skirt) and long-sleeves shirt with clothing insulation of 0.71 clo. Most of the female students wore regular uniform and hijab with clothing insulation 0.80 clo [41]. Hijab is a veil worn by Moslem women, which usually covers the head and chest. Small numbers of female students wore light long-skirt and short-sleeves shirt with clothing insulation 0.67 clo. Except for ensembles with hijab, all other clothing insulation values have been taken or calculated from ASHRAE Standard [36]. The details of clothing ensembles with its clothing insulation using the present study are illustrated in Table 5.

The activities of students during surveys and measurements were mostly sitting and reading. In the ASHRAE standard, this kind of activity has the metabolic rate of 1.0. However, several kinds of literature proposed the estimation of the metabolic rate of students, which is different from an adult. The ISO 8896 provides a guideline in the calculation of metabolic rate [42]. The metabolic rate used in this research has been increased from the figure of adult metabolic rate (1.0 to 1.2) [14,43,44]. The increase of this number is to accommodate the fact that students are smaller than adults, which has smaller body coverage. In fact, this metabolic rate has been used by Wong and Khoo [30] for thermal comfort study of secondary school students in the tropic.

Table 5. The clothing insulation (clo) of ensembles wear by students.

Uniform Type	Ensembles	Clothing Insulation (clo)	Plus Thin Sleeveless Vest (clo)
Men Uniform	Trousers, short-sleeve shirt	0.57	0.67
	Trousers, long-sleeve shirt	0.61	0.71
Moslem Uniform	Trousers, long-sleeve shirt	0.61	-
Sport Uniform	Sweat pants, short-sleeve sweatshirt	0.70	-
Women Uniform	Ankle-length skirt, short-sleeve shirt	0.67	0.77
	Ankle-length skirt, long-sleeve shirt	0.71	0.81
	Ankle-length skirt, long-sleeve shirt, hijab	0.80 *	0.90
Moslem Uniform	Ankle-length skirt, long-sleeve shirt	0.71	-
	Ankle-length skirt, long-sleeve shirt, hijab	0.80 *	-
Sport Uniform	Sweat pants, long-sleeve sweatshirt, hijab	0.84	-

* The value is adopted from [41] all other values are derived from [36].

Figure 2. The situation of survey and measurement in the classrooms. Students clothing for each photo: (**a**) Regular uniform for SMAN; (**b**) Regular uniform for SMPN; (**c**) *Batik* uniform; and (**d**) *Pramuka* uniform.

2.4. Data Processing and Analyses

Data analyses were carried out by spreadsheet software MS Excel and the statistical package for social science (SPSS). The spreadsheet has been used to calculate the mean value of thermal environmental variables and to generate tables showing the microclimatic condition of classrooms. For the statistical analyses, a statistical software SPSS version 16 has been used to calculate regression analyses. The regression analyses examine the correlation and the linearity of data between TSV and operative temperature (T_o), TCV and operative temperature (T_o), and between PMV and operative temperature (T_o). The TSV and TCV were gathered from the respondents' votes written in the questionnaire, while the PMV calculated based on ASHRAE standard [36] using the spreadsheet software developed by Farina [45]. The PMV values calculated by spreadsheet software has a good agreement with calculation example provided in ASHRAE standard [36]. According to Nicol [46], the problem of using PMV in predicting thermal comfort in a hot climate in tropical countries because of the air temperature and the air velocity exceeding the limit that can be handled by PMV. Even though the air temperature exceeds the limit of PMV model, a number of researchers still used it as an indicator to evaluate the thermal comfort of buildings in the tropic [30,32,47]. The TSV votes were grouped according to ASHRAE scale, while the TCV was grouped by using Bedford scale.

To calculate the PMV for each respondent, the availability of four corresponding environmental variables as well as two personal variables for each respondent is essential. However, not all these variables were collected at all points of measurement due to the limitation of equipment. The MRT was only measured at the center point of the classrooms, while air velocity values were only measured at two points, that is, A and B. To simplify the procedure of calculation the MRT values were applied to all points. Regarding air velocity, the arrangements were: the values of air velocity collected in A were shared with the points C and E, while the ones collected at point B were shared with points D and F, respectively. By applying these arrangements, all respondents, who were sitting near the point of measurements, had all six required variables. These enabled us to calculate the PMV values. Based on the four environmental and two personal variables, the PMVs for the 1594 respondents have been calculated.

Research results were analyzed based on the statistical analyses using SPSS version 16. The statistical analysis used in this research was regression analysis, which is based on Pearson correlation. The acceptance of linear regression analyses was determined by two criteria: the test of linearity of regression (*F*-test) and the significance of equation coefficient (*t*-test). The equation is statistically linear if the absolute value of $F > F_{table}$ and *sig.* value < its probability (0.05). The F_{table} for the case is 3.844. The equation coefficients are significant if the absolute value of $t > t_{table}$ and the *sig.* value < half of its probability (0.025 for two tails). The t_{table} for infinity number of degree of freedom is 1.960. The data used for statistical analysis, have been verified by checking their normality and reliability.

3. Results

3.1. Classroom Conditions

The classroom conditions recorded in the 48 classrooms from eight secondary schools. The surveys were conducted in eight days in August 2017 (the dates of the survey are shown in Table 1). Most of the surveys have been carried out in the sunny day where the average outdoor temperature was 29.7 °C, with minimum 23.7 °C in the morning and maximum 33.4 °C in the daytime. The average relative humidity at that time was 69.2% with minimum 50.0% and maximum 85.8%. This outside thermal environment was gathered from Meteorological Station located at the Sultan Hasanuddin International Airport (5°4′ S, 119° 33′ E, 17 m above sea level) about 20 km from city center. The thermal conditions in inside classrooms are presented in Table 6. The table shows that the minimum air temperature 28.2 °C, the average 30.8 °C, and the maximum 33.6 °C. These indicate that these classrooms experienced high temperature during the day. The air temperatures are outside the comfort zone as specified in the national standard [39] and ASHRAE standard [36]. The relative humidity (RH) ranges from 53% to 89% with an average of 68%, indicating that most of the air humidity has already been in the comfort zone. The Indonesian national standard (SNI) 03-6572-2001 specifies that the thermal comfort zone for the comfortable room is within 22.8–25.8 °C T_e (RH about 50%) [39]. Even for the comfortably warm, the air temperatures in the classrooms were not satisfied the SNI, which specifies the comfortably warm zone within 25.98–27.1 °C T_e (RH about 50%). Most of the classes experienced low airflow rate with average velocity was 0.15 m/s.

Table 6. Thermal conditions of the surveyed classrooms.

Microclimatic Factors	Mean	Standard Deviation (SD)	Minimum	Maximum
Air temperature (T_a) in °C	30.8	1.3	28.2	33.6
Relative humidity (RH) (%)	59.8	6.2	44.1	73.2
Mean radian temperature (MRT) in °C	30.8	1.4	28.1	33.6
Operative temperature (T_{op}) in °C	30.8	1.4	28.2	33.6
Air velocity (m/s)	0.15	0.08	0.05	0.50

Figure 3 shows the mean, minimum and maximum operative temperature in the individual school. The mean operative temperature in each surveyed school is very different. Most of the schools have the high mean operative temperature, which in the ranges of 30 to 32 °C. Only two schools have mean operative temperature lower than 30 °C, that is, SMAN 4 and SMAN 1. The figure also indicated that all surveyed schools have hot thermal environments during the school day.

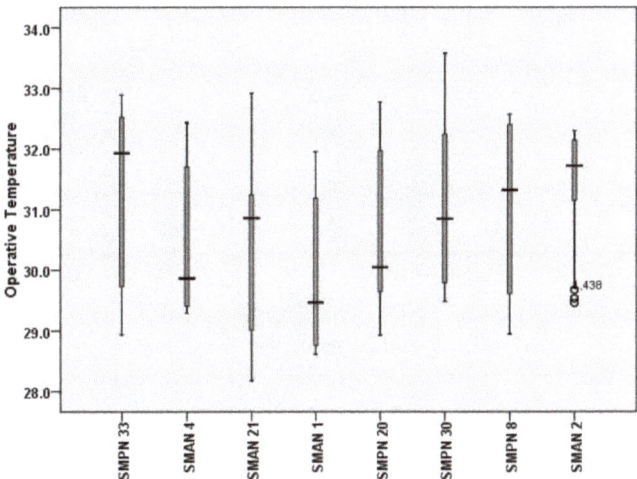

Figure 3. The box plot of operative temperature on each school.

3.2. Students' Responses to the Thermal Environment

Students' response to the thermal environment in the classrooms based on the indicator of thermal sensation votes (TSV) can be seen in Figure 4. As shown in Figure 4, about 37% of respondents voted the neutral (0) option, while about 20% vote slightly cool (−1), and more than 30% vote slightly warm (+1). A total of about 87% students voted for the three central options (−1 to +1). Interestingly, there were only about 12% of them who voted the uncomfortable warm and hot regions (+2 to +3). These votes confirmed that despite the hot temperature in the classrooms, most of the students still felt comfortable.

Regarding the TCV, most students (more than 45%) voted the comfortable (0) option, almost 30% voted comfortably warm (+1), and about 10% of them voted comfortably cool (−1). Similar to the TSV, the percentage of respondents who voted uncomfortable areas (+2 and +3) is 12%. Very few students voted the uncomfortably cold option. Different figures are observed in the predicted votes using the PMV method. In this indicator, only about 23% of respondents were categorized as slightly warm (+1), and more than 77% of respondents felt warm and hot (+2 and +3).

The details comparison of votes between PMV and TSV in different school can be observed in Figure 5. The mean PMV values are lay in between 1.5 and 2.5. The trend of mean PMV is similar to the mean operative temperature of each school as shown in Figure 3. This can be understood that because the PMV is calculated from the operative temperature. Even though the operative temperature quite different from one school to another, the mean TSV votes most of the students from seven schools were very similar, which close to zero (0).

Students' response to the thermal environment in the classrooms based on the indicator of thermal preference can be seen in Figure 6 (left). The figure shows that the majority of respondents (87%) preferred the air temperature in the classrooms to be reduced and that few of them (13%) felt that the temperature was right, so they did not want to increase or decrease the air temperature in the classrooms. Students' response to the thermal environment in the classrooms based on the indicator of thermal acceptance can be seen in the right part of Figure 6. As seen in the figure, the majority of respondents (80%) accepted the conditions of the classrooms; only a small proportion of them (20%) did not accept thermal conditions in their classrooms.

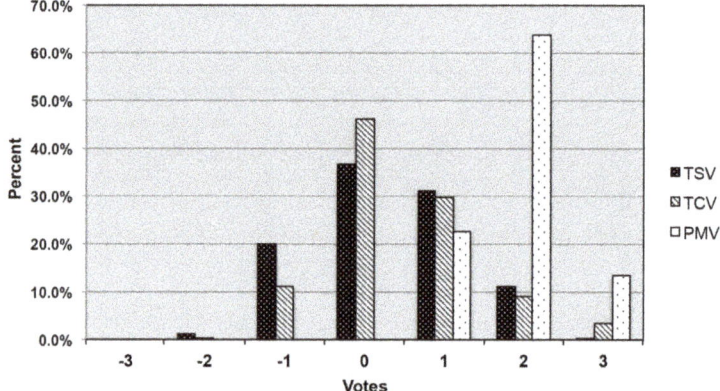

Figure 4. The percentage of PMV, TSV, and TCV.

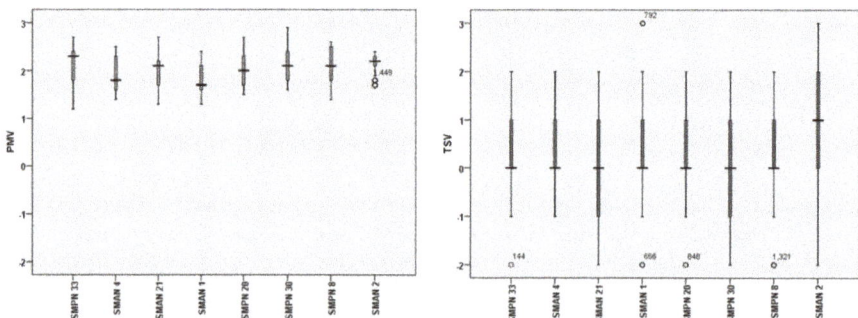

Figure 5. The comparison between the PMV and TSV (thermal sensation votes) of individual schools using boxplot.

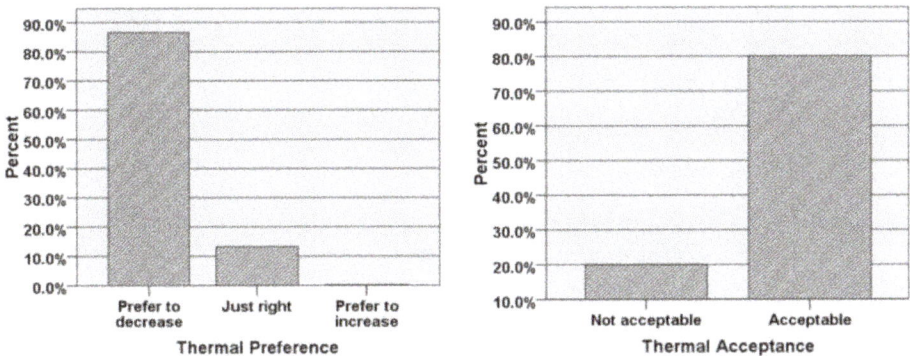

Figure 6. Respondents' thermal preference (**left**) and thermal acceptance (**right**).

3.3. Students' Response to the Air Velocity

Students' response to the airflow in classrooms based on indicators air velocity votes can be seen in Figure 7 (left). The figure shows that the majority of respondents (61%) rarely felt airspeed and only about 31% felt the speed of air flow (just right). These may indicate that the existing natural ventilation system was not able to supply enough airflow into the classrooms. Interestingly, only 1% of respondents felt disturbed (the air flow was too breezy) by the airflow in the classrooms. Students' response to the airflow in classrooms based on air velocity preference can be seen in Figure 7 (right). The figure shows that the majority of respondents (almost 70%) preferred to increase the airspeed and only less than 5% wanted to decrease the speed.

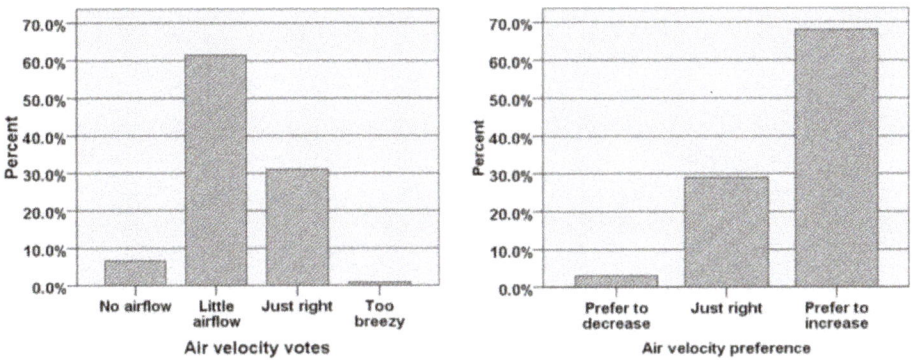

Figure 7. Students' responses to the air velocity (**left**) and their preference (**right**).

3.4. Students' Response to the Air Humidity

Students' responses to the air humidity conditions in the classroom based on humidity votes can be seen in Figure 8. The highest percentage of respondents (41%) felt that the humidity was comfortable (just right). More than 95% of respondents voted at the three scales in the centerline, which indicated that the humidity in the classrooms met the respondents' needs.

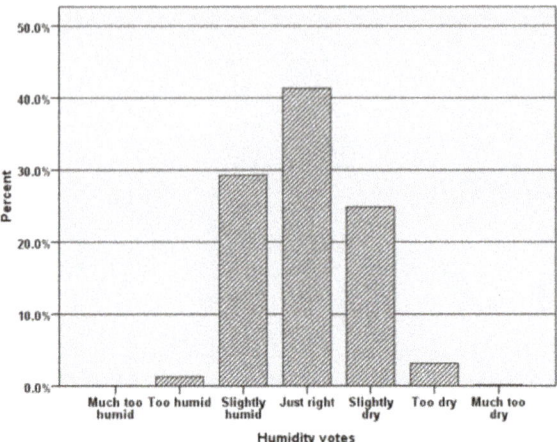

Figure 8. Students' response to the air humidity.

3.5. Neutral Temperature (T_n)

Another indicator for evaluating the thermal comfort is neutral temperature. According to Feriadi and Wong [32], the neutral temperature is the temperature where most of the respondents vote neutral (0) category in the ASHRAE scale. The neutral temperature calculated by applying simple linear regression between the TSV as a dependent variable and the operative temperature (T_o) as an independent variable. For comparison, the authors would also like to include the neutral temperature gathered by the PMV and TCV variables. Figure 9 demonstrates the relationship between the independent variable, operative temperature (T_o) with the dependent variable, PMV, TSV and TCV, respectively. The relationship between these three pairs in producing neutral temperature values is different.

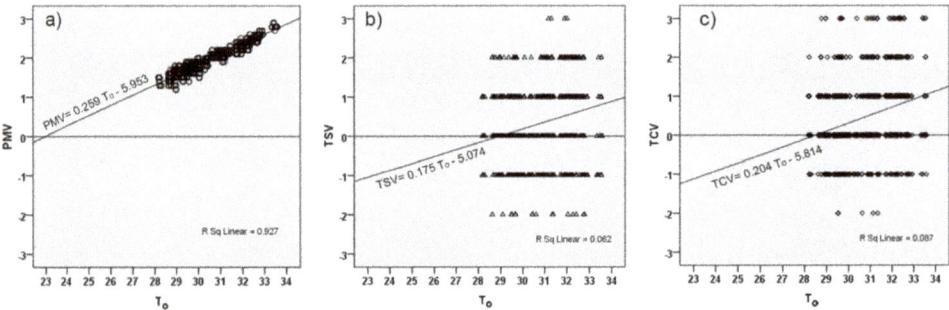

Figure 9. The scatterplot and regression between: (**a**) the operative temperature (T_o) and the PMV; (**b**) he operative temperature (T_o) and the TSV; and (**c**) the operative temperature (T_o) and the TCV (thermal comfort vote).

The first linear regression between the dependent variable PMV and its independent variable (T_o) is presented in Equation (1). The statistical values (R^2 0.927, *F* 20280, and *Sig.* 0.000) indicated that the linear regression between the two variables is statistically significant. The *t*-test for regression coefficients showed that the *t* values of constant ($|-106.094|$) and the operative temperature (142.406) are higher than 1.960, which indicated that both coefficients are statistically significant. By using

Equation (1), the value of PMV = 0 (T_n) will be obtained when the operative temperature is 23.0 °C, which means that the respondents would feel neutral at an operative temperature of 23.0 °C.

$$PMV = 0.259T_o - 5.953, \tag{1}$$

The second linear regression between the dependent variable (TSV) and its independent variable (T_o) is shown in Equation (2). The linear regression is statistically significant. This is supported by the statistical values of R^2 0.062, F 104.369, and *Sig.* 0.000. The *t*-test for regression coefficients showed that the *t* values of coefficient of constant ($|-9.604|$) and the operative temperature (10.216) are higher than 1.960, which indicated that both coefficients are statistically significant. Therefore the Equation (2) can be used to predict the TSV when the T_o is known. By using Equation (2), the value of TSV = 0 (T_n) will be obtained when the operative temperature is 29.0 °C, which means that the respondents would feel neutral at an operative temperature around 29.0 °C. With the operative temperature of classrooms ranging from 28.2 °C up to 33.6 °C, some respondents would predictably feel neutral.

$$TSV = 0.175T_o - 5.074, \tag{2}$$

The third linear regression between the dependent variable (TCV), and its independent variable (T_o) is presented in Equation (3). The statistical values R^2 0.087, F 152.381, and *Sig.* 0.000 indicated that the linear regression between the two variables is statistically significant. The *t*-test for regression coefficients showed that the *t* values of coefficient of constant ($|-11.415|$) and the operative temperature (12.344) are higher than 1.960, which indicated that both coefficients are statistically significant. Therefore the Equation (3) can be used to predict the TCV when the T_o is known. By using Equation (3), the value of TCV = 0 (T_n) will be obtained when the operative temperature is 28.5 °C. This means that the respondents would feel comfortable at an operative temperature about 28.5 °C. This value is similar to the neutral temperature obtained by Equation (2).

$$TCV = 0.204T_o - 5.814 \tag{3}$$

4. Discussion

The classroom condition of the surveyed secondary schools shows hot thermal environments. The indoor air temperature ranging from 28.2 °C to 33.6 °C with average 30.8 °C, which has already beyond the thermal comfort zone as specified in the ASHRAE 55 standard [36] and the national standard [39]. This condition could have the problem for students [48]. However, the students' responses on the thermal sensation give very different figures, where more than 80% voted within the central (−1 to +1) option. This thermal environment quite similar to the one recorded in the naturally ventilated secondary school in Singapore [30] and in the residential buildings in Jogyakarta [32]. These indicate that naturally ventilated buildings in the tropic area in South East Asia experience hot temperature during daytime.

The high percentage of students (more than 85%) voted in the centerline options (−1 to +1) either in TSV of TCV shows that students are comfortable in the classrooms despite their hot temperature. The percentage of respondents who voted 0 (comfortable) in the TCV is higher than the percentage of respondents who voted 0 (neutral) in the TSV. A similar result was found in Feriadi and Wong [32], where about 40% of respondents voted the 0 (comfortable) in the TCV, and only less than 20% voted the 0 (neutral) in TSV. Feriadi and Wong [32] proposed two reasons for the difference between TSV and TCV. Firstly, respondents are more stringent when they vote for the thermal sensation than the thermal comfort perception. Secondly, there is a tendency for occupants in naturally ventilated buildings to perceive cold (coolness) as comfortable.

The mean vote of PMV model overestimates the actual votes of respondents in TSV and TCV. This result agrees with a result of a study done by Feriadi and Wong [32], where they found that the PMV always overestimated the actual votes of respondents in the residential buildings in the tropics.

More respondents are predicted to vote uncomfortably warm and hot in PMV rather than in the TSV and TCV. To improve the accuracy of PMV in predicting the thermal sensation of occupants Fanger and Toftum [49] propose an extension of PMV to be applicable in the naturally ventilated buildings. In this study, Fanger and Toftum proposed the reduction of the estimated metabolic rate, and the calculation of expectation factor e. The expectation factor derived from the equation as follows:

$$TSV = e \times PMV, \qquad (4)$$

The expectation factor e gathered from Equation (4) is 0.15. This figure is very low in comparison to the lowest expectation factor proposed by Fanger and Toftum [49], with a minimum value of 0.5. The expectation factor e 0.5 is suitable for non-air conditioned buildings located in regions with few air conditioned buildings, warm weather during all seasons [49].

We already know that one reason for the discrepancy between the PMV and the actual votes TSV and TCV is that the accuracy of data inputted into the model. We can argue that disagreement between PMV and TSV is mostly dependent on the determination of clothing insulation and metabolic rate used in the PMV model because the four other variables related to the thermal environment were recorded directly from the surveyed classrooms. For example, using the clo values as specified in the ASHRAE standard [36] will give higher PMV value in comparison to the clo values used in the Teli et al. [14]. This is also true for the determination of metabolic rate. This has a big impact on the calculation of PMV. Therefore, the clothing insulation and metabolic rate should be calculated accurately as demonstrated in Al-ajmi et al. [41], Havenith [43], and Haddad et al. [44].

The neutral temperatures obtained the actual votes (TSV and TCV) are very different with the neutral temperature obtained the predicted one (PMV). The PMV predicts the neutral temperature for this case is about 23 °C, which about 6 °C lower than the TSV and TCV. In the room with operative temperatures (T_o) ranging from 28.2 °C to 33.6 °C, no respondents will be predicted neutral or comfortable using this PMV model. The neutral temperature gathered from actual votes TSV is 29.0 °C. This temperature has a good agreement with studies carried out in the tropics. For example, neutral temperature found by Wong and Khoo [30] in school buildings in the tropical city of Singapore is only 0.2 °C lower than this present study. Based on the two days survey and measurements in the secondary school, they found the neutral temperature of 28.8 °C. Another study by de Dear [47] found the neutral temperature of 28.5 °C. Feriadi and Wong [32] found the neutral temperature in Jogyakarta Indonesia of about 29.2 °C, which is a little bit higher than the neutral temperature found in this study.

Considering the limitation of applicability of PMV model in predicting the thermal sensation and neutral temperature of respondents, the adaptable thermal comfort should be used. Several schools have provided means of increasing thermal comfort such as fan and openable windows that can be open when in need. Some female students also carried a small battery-operable portable fan to increase her comfort. The explanation of this adaptable thermal comfort will be discussed in another article in the future, which is combined the survey carried out in the primary and secondary schools.

5. Conclusions

The measurement of classrooms in the eight selected secondary schools in the tropical city of Makassar showed hot air environments. The air temperatures ranged from 28.2 °C in the morning to 33.6 °C in the midday. The radiant temperatures (28.2 °C to 33.6 °C) were very similar to the air temperature. The airflow speeds were fairly stagnant, characterized by an average of 0.15 m/s and a maximum of 0.5 m/s. The only parameter that could meet the Indonesia national standard (SNI) is the air humidity, which is ranging from 44% to 73% with an average of 60%.

The thermal comfort survey results show that secondary school students were tolerant of the hot temperatures. About 37% of respondents voted the neutral (0) option, while about 20% voted slightly cool (−1), and more than 30% voted slightly warm (+1). Interestingly, there were only about 12% of them who voted the warm and hot regions (+2 to +3). These votes confirmed that despite the hot temperature in the classrooms, most of the students still felt comfortable. In fact, more than 86% of respondents accepted these thermal conditions. Even though the acceptance rate is very high, if there is a chance, 72% of respondents wanted cooler air temperature. Most of the respondents did not feel the airflow in the classrooms so that most wanted the increase in airflow velocity. A very small portion of respondents complained about the air humidity.

The result of the calculation of neutral temperature (T_n) using a model predicted mean vote (PMV) produced the value 23.0 °C which is very low when compared with the operative temperature in the room. It is smaller than the value of the neutral temperature (T_n) obtained from the actual votes either by TSV or TCV. The T_n obtained from TSV and TCV were 29.0 °C and 28.5 °C, respectively.

This study suggests that in the tropical Indonesian city, the secondary school students are able to acclimatize themselves with respect to thermal environments, which are beyond the comfort zone specified by the international and national standards. The evidence encourages the use of passive design in the school building construction and operation. There is a great potential for passive design working with electrical fans to lower the classroom temperature in response to students' needs. This study, on the other hand, suffers from several limitations, especially the availability of equipment, which handicapped the accuracy of the measurement. Future studies shall deploy more meters and sensors working simultaneously to collect data on thermal environments.

Acknowledgments: We would like to express our sincere gratitude and appreciation to the Minister of Research, Technology and Higher Education and the Rector of Hasanuddin University for providing the fund for this research, with contract number 005/SP2H/LT/DRPM/IV/2017, Dated 23 April 2017. We would also like to thank all doctoral students at Laboratory of Building Science and Technology, the Department of Architecture, Hasanuddin University, especially to Mr. Sahabuddin, Mr. Ansarullah, Mr. Tayeb Mustamin, and Ms. Irnawaty Idrus, who involved in the data collections. Lastly, we would like to acknowledge the help made by Mr. Muhammad Faizal for collecting and processing the data gathered from the surveys.

Author Contributions: All authors contributed equally to this paper.

Conflicts of Interest: The authors declare no conflict of interest.

References

1. ASHRAE. *Thermal Comfort Conditions (Standard 55-66)*; ASHRAE: New York, NY, USA, 1966.
2. ISO. *Moderate Thermal Environment—Determination of the PMV and PPD Indices and Specifications for Thermal Comfort*, 2nd ed.; ISO 7730; International Standard Organisation (ISO): Geneva, Switzerland, 1995.
3. Gagge, A.P.; Stolwijk, J.A.J.; Hardy, J.D. Comfort and thermal sensations and associated physiological responses at various ambient temperatures. *Environ. Res.* **1967**, *1*, 1–20. [CrossRef]
4. Fanger, P.O. *Thermal Comfort—Analysis and Application in Environmental Engineering*; Danish Technical Press: Copenhagen, Denmark, 1970.
5. D'Ambrosio Alfano, F.R.; Olesen, B.W.; Palella, B.I. Povl Ole Fanger's impact ten years later. *Energy Build.* **2017**, *152*, 243–249. [CrossRef]
6. Gou, Z.; Lau, S.S.-Y.; Chen, F. Subjective and Objective Evaluation of the Thermal Environment in a Three-Star Green Office Building in China. *Indoor Built Environ.* **2012**, *21*, 412–422. [CrossRef]
7. Gou, Z.; Lau, S.-S.; Lin, P. Understanding domestic air-conditioning use behaviours: Disciplined body and frugal life. *Habitat Int.* **2017**, *60*, 50–57. [CrossRef]
8. Sensharma, N.P.; Woods, J.E.; Goodwin, A.K. Relationship between the indoor environment and productivity: A literature review. *ASHRAE Trans.* **1998**, *104*, 686–701.
9. Seppänen, O.; Fisk, W.J.; Lei-Gomez, Q. Effect of Temperature on Task Performance in Offfice Environment. In Proceedings of the 5th International Conference on Cold Climate Heating, Ventilating and Air Conditioning, Moscow, Russia, 21–24 May 2006; p. 53-16.
10. Al horr, Y.; Arif, M.; Katafygiotou, M.; Mazroei, A.; Kaushik, A.; Elsarrag, E. Impact of indoor environmental quality on occupant well-being and comfort: A review of the literature. *Int. J. Sustain. Built Environ.* **2016**, *5*, 1–11. [CrossRef]
11. Pepler, R.D.; Warner, R.E. Temperature and learning: An experimental study. *ASHRAE Trans.* **1968**, *74*, 211–224.

12. Bellia, L.; Boerstra, A.; da Silva, M.C.G.; Ianniello, E.; Lopardo, G.; Minichiello, F.; Romagnoni, P.; van Dijken, F. *Indoor Environment and Energy Efficiency in Schools—Part 1*; REHVA: Brussels, Belgium, 2010.
13. Mendell, M.J.; Heath, G.A. Do indoor pollutants and thermal conditions in schools influence student performance? A critical review of the literature. *Indoor Air* **2005**, *15*, 27–52. [CrossRef] [PubMed]
14. Teli, D.; Jentsch, M.F.; James, P.A.B. Naturally ventilated classrooms: An assessment of existing comfort models for predicting the thermal sensation and preference of primary school children. *Energy Build.* **2012**, *53*, 166–182. [CrossRef]
15. Teli, D.; Jentsch, M.F.; James, P.A.B. The role of a building's thermal properties on pupils' thermal comfort in junior school classrooms as determined in field studies. *Build. Environ.* **2014**, *82*, 640–654. [CrossRef]
16. Teli, D.; Bourikas, L.; James, P.A.B.; Bahaj, A.S. Thermal Performance Evaluation of School Buildings using a Children-based Adaptive Comfort Model. *Procedia Environ. Sci.* **2017**, *38*, 844–851. [CrossRef]
17. Buratti, C.; Ricciardi, P. Adaptive analysis of thermal comfort in university classrooms: Correlation between experimental data and mathematical models. *Build. Environ.* **2009**, *44*, 674–687. [CrossRef]
18. Corgnati, S.P.; Filippi, M.; Viazzo, S. Perception of the thermal environment in high school and university classrooms: Subjective preferences and thermal comfort. *Build. Environ.* **2007**, *42*, 951–959. [CrossRef]
19. Corgnati, S.P.; Ansaldi, R.; Filippi, M. Thermal comfort in Italian classrooms under free running conditions during mid seasons: Assessment through objective and subjective approaches. *Build. Environ.* **2009**, *44*, 785–792. [CrossRef]
20. Alfano, F.R.D.A.; Palella, B.I.; Riccio, G. Notes on the Calculation of the PMV Index by Means of Apps. *Energy Procedia* **2016**, *101*, 249–256. [CrossRef]
21. D'Ambrosio Alfano, F.R.; Ianniello, E.; Palella, B.I. PMV–PPD and acceptability in naturally ventilated schools. *Build. Environ.* **2013**, *67*, 129–137. [CrossRef]
22. Huang, K.-T.; Huang, W.-P.; Lin, T.-P.; Hwang, R.-L. Implementation of green building specification credits for better thermal conditions in naturally ventilated school buildings. *Build. Environ.* **2015**, *86*, 141–150. [CrossRef]
23. Hwang, R.-L.; Lin, T.-P.; Kuo, N.-J. Field experiments on thermal comfort in campus classrooms in Taiwan. *Energy Build.* **2006**, *38*, 53–62. [CrossRef]
24. Hwang, R.L.; Lin, T.P.; Chen, C.P.; Kuo, N.J. Investigating the adaptive model of thermal comfort for naturally ventilated school buildings in Taiwan. *Int. J. Biometeorol.* **2009**, *53*, 189–200. [CrossRef] [PubMed]
25. Liang, H.-H.; Lin, T.-P.; Hwang, R.-L. Linking occupants' thermal perception and building thermal performance in naturally ventilated school buildings. *Appl. Energy* **2012**, *94*, 355–363. [CrossRef]
26. Zaki, S.A.; Damiati, S.A.; Rijal, H.B.; Hagishima, A.; Razak, A.A. Adaptive thermal comfort in university classrooms in Malaysia and Japan. *Build. Environ.* **2017**, *122*, 294–306. [CrossRef]
27. Kwok, A.G.; Chun, C. Thermal comfort in Japanese schools. *Sol. Energy* **2003**, *74*, 245–252. [CrossRef]
28. De Dear, R.; Kim, J.; Candido, C.; Deuble, M. Adaptive thermal comfort in Australian school classrooms. *Build. Res. Inf.* **2015**, *43*, 383–398. [CrossRef]
29. Kim, J.; de Dear, R. Thermal comfort expectations and adaptive behavioural characteristics of primary and secondary school students. *Build. Environ.* **2018**, *127*, 13–22. [CrossRef]
30. Wong, N.H.; Khoo, S.S. Thermal comfort in classrooms in the tropics. *Energy Build.* **2003**, *35*, 337–351. [CrossRef]
31. Kwok, A.G. Thermal comfort in tropical classrooms. *ASHRAE Trans.* **1998**, *104*, 1031–1047.
32. Feriadi, H.; Wong, N.H. Thermal comfort for naturally ventilated houses in Indonesia. *Energy Build.* **2004**, *36*, 614–626. [CrossRef]
33. Karyono, T.H. Report on thermal comfort and building energy studies in Jakarta—Indonesia. *Build. Environ.* **2000**, *35*, 77–90. [CrossRef]
34. Hamzah, B.; Ishak, M.T.; Beddu, S.; Osman, M.Y. Thermal comfort analyses of naturally ventilated university classrooms. *Struct. Surv.* **2016**, *34*, 427–445. [CrossRef]
35. Karyono, T.H.; Heryanto, S.; Faridah, I. Air conditioning and the neutral temperature of the Indonesian university students. *Archit. Sci. Rev.* **2015**, *58*, 174–183. [CrossRef]
36. ASHRAE. *Thermal Environmental Conditions for Human Occupancy (ANSI/ASHRAE Standard 55-2013)*; ASHRAE: Atlanta, GA, USA, 2013.
37. Bedford, T. *Warmth Factor in Comfort at Work*; Medical Res Council, Industry Health Res Board: London, UK, 1936.

38.	Soegijanto. *Bangunan di Indonesia dengan Iklim Tropis Lembab Ditinjau dari Aspek Fisika Bangunan (Buildings in Indonesia with the Humid Tropical Climate Considering Building Physics Aspects)*; Direktorat Jenderal Pendidikan Tinggi Departemen Pendidikan dan Kebudayaan: Jakarta, Indonesia, 1999.

39.	BSN. *SNI 03-6572-2001: Tata Cara Perancangan Sistem Ventilasi dan Pengkondisian Udara Pada Bangunan Gedung (The Procedure of Designing Ventilation and Air Conditioning Systems in Buildings)*; Badan Standardisasi Nasional (BSN): Jakarta, Indonesia, 2001.

40.	Auliciems, A.; Szokolay, S.V. *PLEA Notes: Note 3 Thermal Comfort*; PLEA: Passive and Low Energy Architecture International in association with Department of Architecture, The University of Queensland: Brisbane, Australia, 2007.

41.	Al-ajmi, F.F.; Loveday, D.L.; Bedwell, K.H.; Havenith, G. Thermal insulation and clothing area factors of typical Arabian Gulf clothing ensembles for males and females: Measurements using thermal manikins. *Appl. Ergon.* **2008**, *39*, 407–414. [CrossRef] [PubMed]

42.	ISO. *Ergonomics of the Thermal Environment—Determination of Metabolic Rate*; ISO 8996:2004(E); International Standardization Organization: Geneva, Switzerland, 2004.

43.	Havenith, G. Metabolic rate and clothing insulation data of children and adolescents during various school activities. *Ergonomics* **2007**, *50*, 1689–1701. [CrossRef] [PubMed]

44.	Haddad, S.; Osmond, P.; King, S. Metabolic rate estimation in the calculation of the PMV for children. In *Cutting Edge in Architectural Science, Proceedings of the 47th International Conference of the Architectural Science Association, Hong Kong, China, 13–16 November 2013*; The Architectural Science Association: Hong Kong, China, 2013.

45.	Farina, A. PMV Calculator. 2015. Available online: http://www.angelofarina.it/Public/Fisica-Tecnica-Ambientale-2015/Lez-04-05/PMV_cal_tanabe6.xls (accessed on 10 December 2015).

46.	Nicol, F. Adaptive thermal comfort standards in the hot–humid tropics. *Energy Build.* **2004**, *36*, 628–637. [CrossRef]

47.	De Dear, R.J.; Leow, K.G.; Ameen, A. Thermal comfort in the humid tropics. Part 1 Climate chamber experiments on temperature preferences in Singapore. *ASHRAE Trans.* **1991**, *97*, 874–879.

48.	Puteh, M.; Ibrahim, M.H.; Adnan, M.; Che'Ahmad, C.N.; Noh, N.M. Thermal Comfort in Classroom: Constraints and Issues. *Procedia Soc. Behav. Sci.* **2012**, *46*, 1834–1838. [CrossRef]

49.	Fanger, P.O.; Toftum, J. Extension of the PMV model to non-air-conditioned buildings in warm climates. *Energy Build.* **2002**, *34*, 533–536. [CrossRef]

 © 2018 by the authors. Licensee MDPI, Basel, Switzerland. This article is an open access article distributed under the terms and conditions of the Creative Commons Attribution (CC BY) license (http://creativecommons.org/licenses/by/4.0/).

Article

An Investigation of Thermal Comfort and Adaptive Behaviors in Naturally Ventilated Residential Buildings in Tropical Climates: A Pilot Study

Zhonghua Gou [1,*], Wajishani Gamage [2], Stephen Siu-Yu Lau [3] and Sunnie Sing-Yeung Lau [4]

[1]   School of Engineering and Built Environment, Griffith University, Gold Coast, QLD 4215, Australia
[2]   Department of Architecture, University of Moratuwa, Moratuwa 10400, Sri Lanka; waji724@gmail.com
[3]   Department of Architecture, National University of Singapore, Singapore 117566, Singapore;
      akilssy@nus.edu.sg
[4]   Department of Architecture, University of Hong Kong, Hong Kong, China; sunnielau@gmail.com
*   Correspondence: z.gou@griffith.edu.au or gouzhonghua@gmail.com; Tel.: +61-7-5552-9510

Received: 9 November 2017; Accepted: 2 January 2018; Published: 3 January 2018

Abstract: This article presents a pilot study of thermal comfort and adaptive behaviors of occupants who live in naturally ventilated dormitories at the campus of the National University of Singapore. A longitudinal survey and field measurement were conducted to measure thermal comfort, adaptive behaviors and indoor environment qualities. This study revealed that occupants living in naturally ventilated buildings in tropics were exposed to higher operative temperatures than what American Society of Heating, Refrigerating and Air-Conditioning Engineers (ASHRAE) comfort standards recommend for naturally conditioned spaces. However, they still felt that such conditions were acceptable. Two behavioral adjustments were found to have profound impacts on occupants' acceptance of the imposed heat stresses: (1) increasing the indoor air velocity by turning on mechanical fans and opening the door/windows for cross ventilation, and (2) reducing clothing insulation by changing clothes and dressing in fewer clothes. Higher indoor air velocities were also associated with greater satisfaction with indoor air quality. The future study should develop a statistical model to correlate adaptive behaviors with temperature variations for tropical climates.

Keywords: natural ventilation; thermal comfort; thermal adaption; tropical climate; indoor environment quality

---

1. Introduction

The construction and operation of buildings accounts for over 70% of global greenhouse gas emissions [1]. A significant portion of the end-use energy in buildings is dedicated to maintaining thermal comfort [2]. In tropical climates, buildings are exposed to high solar radiation throughout the whole year; therefore, these buildings are greatly dependent on air-conditioning to provide occupants with comfortable thermal environments [3,4]. Singapore's buildings are responsible for 37% of the country's total electricity consumption; among all consumptions, air conditioning systems are responsible for 40–50% of this total energy consumption [5]. Singapore's continuous efforts on the adoption of green building design and technologies had resulted in an improvement of their energy utilization index (EUI) by 7% from 2008 to 2014 [5]. This has been attributed to energy-efficient improvements in buildings. The Building Construction Authority (BCA) of Singapore advocates natural ventilation as a priority in green and sustainable design [6].

Thermal comfort is the condition of mind that expresses satisfaction with the thermal environment [7]. The thermal comfort depends on many factors beyond air temperature, such as air velocity, relative humidity, season, clothing insulation, activity level and others. Generally, the American

Society of Heating, Refrigerating and Air-Conditioning Engineers (ASHRAE) recommends an indoor air temperature range of 19–28 degrees Celsius for thermal comfort purposes [7]. The upper relative humidity level recommended by the standard is 80% [7]. The climate of Singapore is characterized by a high ambient temperature range of 26–33 degrees Celsius and relative humidity of 70–90% throughout the year (Table 1) [8]. According to ASHRAE Standard 55, "Thermal Environmental Conditions for Human Occupancy" [9], cooling and dehumidification is required for 76.6% of the hours from January to December for thermal comfort in Singapore [10]. Under such circumstances, the application of natural ventilation for indoor comfort is uncertain.

Table 1. The climate of Singapore (1984–2015) (Source: Meteorological Service Singapore [8]).

	Month	Jan.	Feb.	Mar.	Apr.	May	Jun.	Jul.	Aug.	Sep.	Oct.	Nov.	Dec.	Annual
	Mean Daily Maximum	30.1	31.1	31.6	31.8	31.7	31.3	30.9	30.9	30.9	31.1	30.7	30.0	31.0
Temperature (°C)	Mean Daily Minimum	23.3	23.6	24.0	24.4	24.8	24.8	24.6	24.5	24.3	24.1	23.8	23.5	24.1
	24-h Mean	26.0	26.5	27.0	27.4	27.7	27.8	27.5	27.3	27.2	27.1	26.5	26.0	27.0
	Mean Daily Maximum	95.7	95.5	96.2	96.8	96.3	95.3	94.8	94.9	95.7	96.3	97.1	96.7	95.9
Relative Humidity (%)	Mean Daily Minimum	65.5	61.2	61.9	63.2	64.6	64.0	64.8	64.5	63.7	63.0	65.9	68.2	64.2
	24-h Mean	84.6	82.7	83.6	84.6	84.2	82.7	82.6	82.8	83.2	83.8	86.2	86.8	84.0
Wind Speed (m/s)	Mean Monthly/Annual	2.6	2.8	2.1	1.5	1.6	2.0	2.3	2.4	2.0	1.5	1.4	2.0	2.0

Most Singaporeans (86% of the population) live in Housing Developments (HDB) residential buildings which are designed for natural ventilation. The statistical data for energy usage in Singapore show that the domestic energy consumption has doubled since the year 1991 [11]. The air-conditioning system saturation rate (the percentage of households that had air-conditioners) in the residential building sector was 19.4% in 1988 [12]; whereas the figure raised to 88% in the year of 2015 [13]. As a leading research institute and a testbed for green initiatives, the National University of Singapore (NUS) announced their commitment to reduce carbon emissions by 23% by the year 2020 [14]. Reduction of emissions related to electricity consumption is a key initiative to decrease NUS carbon emissions. Air conditioning accounts for 60% of the overall building energy consumption at NUS [14]. To reduce the excessive use of air conditioning, NUS has taken measures to increase indoor temperature setting-point to 25 degrees Celsius. Furthermore, a large portion of student dormitories are designed for natural ventilation without air-conditioning. The NUS dormitories without air-conditioning represent unique opportunities for the study of thermal comfort and adaptive behaviors in tropical climates.

2. Natural Ventilation and Thermal Adaption

A naturally-ventilated residential building aims to provide comfortable indoor thermal environments for residents without using air-conditioning. It is well known that the acclimatization to tropical climates is different from that to cold or temperate climates [15,16]. Many studies indicated that people living or working in naturally ventilated buildings in warm climates accepted high temperature and humidity [17–21]. Field studies on thermal comfort conducted in naturally ventilated buildings in Singapore indicated that ASHRAE standards might not be suitable to predict thermal comfort in free-running buildings in the local climate [22,23]. Thermal perceptions and tolerance in naturally ventilated buildings are likely to make occupants accept a wider range of temperatures as a result of thermal adaptation. Thermal adaptation to relatively moderate fluctuations within free running buildings can be mainly attributed to behavioral adjustments and psychological adaptation [24]. Behavioral adjustments due to a person's conscious or unconscious modifications result in changes to the heat and mass fluxes governing the body's thermal balance [25]. Behavioral adjustments include modification made to one self (clothing, activity, eating or drinking) and modifications to the surrounding (opening a window, turning fans on etc.). Behavioral adjustment of the body's heat balance offers the greatest opportunity for people to actively maintain their comfort. Behavioral adjustments such as changes to clothing insulation may even lead to a 3 degrees Celsius offset in preferred temperatures [26]. Psychological adaptation accounts for the cognitive and cultural variables which affect one's habituation and expectation. The adaptive model to thermal comfort acknowledges a potential feedback loop where past and present thermal experiences with the indoor and outdoor environment directly affect thermal response and cognitive assessment [24].

The adaptive approach to thermal comfort allows for the moderation of the narrow thermal environments specified by the static heat balance model [27]. The adaptive thermal comfort standard is generally applied to naturally conditioned spaces where no mechanical cooling system is installed and the mean outdoor temperatures vary between 10–33 degrees Celsius [7]. Adaptive thermal comfort can be expected for 5882 h annually (67.1%) in tropical climates such as Singapore through natural ventilation [10]. A significant reduction of energy required for indoor comfort can be achieved by understanding the comfort expectations and behavior of occupants. In residential buildings, occupants have greater opportunities to adapt to the thermal environment, which indicates an opportunity to reduce energy consumption. The adaptive behaviors haven been studied in temperate climates, where Andersen et al. [28] conducted repeated surveys of occupant control of the indoor environment in Danish dwellings. They carried out multiple logistic regression to quantify factors influencing occupants' behaviors regarding the factors of windows open/closed, heating on/off, lighting on/off and solar shading in/not in use. It was found that window opening behavior was strongly related to the outdoor temperature. Window opening behavior was also affected by the perception of the environment and factors concerning the dwelling. Kim et al. [29] examined the pattern of adaptive comfort behavior in residential units with both natural ventilation and air-conditioning in Sydney which belongs to subtropical climates. They derived statistical models to predict the percentage of adaptive behaviors such as operation of air-conditioners, heaters, fans and windows, as a function of outdoor temperature. The adaptive behaviors and related thermal environments can also be found in office buildings with different climates [30–32].

According to the theory of adaptive thermal comfort, building occupants play an active role to achieve comfort by adjusting their behaviors. It is unnecessary to maintain a narrow range of temperature using mechanical solutions, especially considering the large amount of energy used to maintain the narrow comfort zone. The adaptive approach of understanding thermal comfort can help to design buildings with passive strategies while reduce the reliance on mechanical solutions, which in turn help to reduce the energy consumption due to cooling and heating. This is extremely important for sustainable building design in tropical climates, where cooling accounts for a large portion of building energy use. There is strong research need to understand occupants' behaviors in naturally ventilated buildings in tropical climates. Focusing on naturally ventilated residential buildings in tropical climates, the aim of this research is to investigate occupants' perceptions and adaptation to heat stresses in these buildings, based on which passive design strategies can be explored.

3. Methodology

A field survey was conducted within two university dormitory buildings: A & B, located at the National University of Singapore, as shown in Figure 1. As a pilot study, the survey took place in six student rooms. All participants were university students with ages between 20 and 30. The layout plan, orientation and location of indoor environmental quality monitoring equipment in each room are shown in Figure 2. Room 1, 2 and 3 are located on the 4th floor of Building A; Room 4 and 5 are located on the 20th floor of Building B; and Room 6 is located on the 18th floor of Building B. All the surveyed rooms were equipped with ceiling fans, operable windows, and window blinds. In some rooms, the occupants had bought their own portable electrical fans. The rooms in Building A are directly connected to a common corridor which is naturally ventilated. The rooms in Building B are connected to the common living area which is naturally ventilated. Furthermore, these rooms have an adjustable air vent on the wall (Figure 3).

Figure 1. Surveyed buildings: Building-A (**a**) and Building-B (**b**).

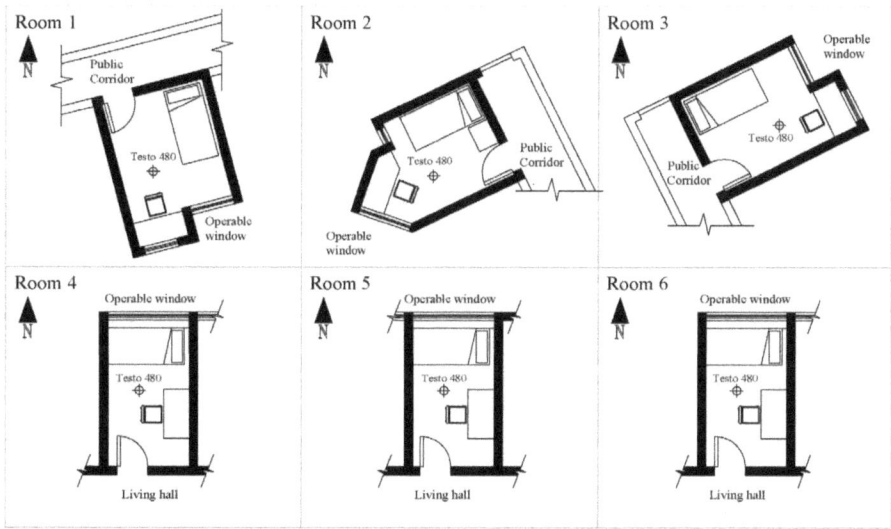

Figure 2. Floor plans and locations of the measured point in each student room.

 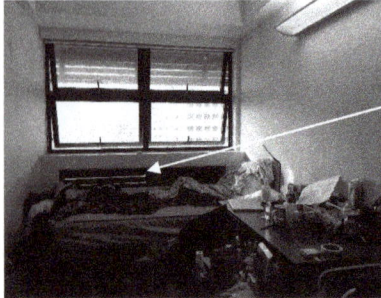

Adjustable air vent on walls

Figure 3. Two typical settings of surveyed student rooms.

A longitudinal thermal comfort survey was carried out in two phases which spanned from 24 May to 2 August 2016. In both phases, the occupants filled out an online questionnaire on indoor environment qualities, adaptive behaviors and the state of indoor environment controls. Phase 1 spanned for two weeks, where the occupant filled in an online questionnaire three times per day (morning, afternoon and evening). The air temperature (Ta) and relative humidly (RH) inside the rooms were measured using HOBO data loggers U12-012 (Onset, Bourne, MA, USA) at 30 s intervals for two consecutive weeks. All measurements were taken at 0.75 m height above floor. Phase 2 spanned Phase 1's daily survey for nine consecutive hours from 10:00 a.m. to 06:00 p.m., where the occupant filled in the online questionnaire every hour. The air temperature (Ta), relative humidity (RH), globe temperature (Tg) and air speed (Va) inside the rooms were measured within 1 m distance from the occupant at a height of 1.1 m above the floor. The data was logged using TESTO 480 (Testo, Lenzkirch, Germany) digital indoor environment quality meter at 1 minute intervals. Both mean radiant temperature (Tmrt) and operative temperature (Top) were calculated in reference to ASHRAE Standard 55 [7].

The online questionnaire was adapted from the following surveys: CBE (The Center for Built Environment) Occupant IEQ Survey [33] and Building Use Studies (BUS) Occupant Survey [34,35]. The questionnaire was prepared in English and was distributed to the occupants via email; the response of each occupant was monitored online through the online platform Google Forms. As shown in Table 2, the questionnaire consisted of six main sections: thermal comfort, air quality, lighting quality, noise, behavioral adjustments and the state of indoor environment controls (i.e., ceiling fans, position of door, windows and blinds). Thermal comfort was measured using the standard seven-point ASHRAE thermal sensation scale, thermal acceptability scale and thermal preferences [7]. Occupant's acceptance and perceptions towards different aspects of the air quality, lighting quality and acoustics were measured on a seven-point scale. The occupants were also asked to choose adaptive behaviors from a list of typical behaviors such as consumption of drinks, adjustments to clothing and operation of ventilation systems controls. They listed all adaptive behaviors and the most frequently performed adaptive behaviors one hour prior to each response. The occupants further rated the accessibility, usability and responsiveness of their most frequent adaptive behaviors on a seven-point scale. The questionnaire also recorded clothing insulation and activity at the time of response. A total of 225 valid responses were received during Phase 1, and 54 valid responses were received during Phase 2. The data from the questionnaire survey were analyzed using statistical software IBM SPSS Statistics 23.0 provided by the National University of Singapore.

Table 2. The questionnaire items.

Question	Abbreviation	Scale Point	
		1	7
Thermal comfort	TC	Very acceptable	Not at all acceptable
Thermal sensation vote	TSV	Cold	Hot
Air movement	AM	Very acceptable	Not at all acceptable
Air quality	AQ	Very acceptable	Not at all acceptable
Perceived draft	DRAFT	Still	Drafty
Perceived humidity	HUMID	Dry	Humid
Air freshness	AF	Fresh	Stuffy
Overall lighting	LQ	Very acceptable	Not at all acceptable
Natural light amount	NLA	Too little	Too much
Natural light glare	NLG	None	Too much
Artificial light amount	ALA	Too little	Too much
Artificial light glare	ALG	None	Too much
Noise level acceptance	NOISE	Very acceptable	Not at all acceptable
Most frequent adaptive action—Accessibility	FAA	Very convenient	Very inconvenient
Most frequent adaptive action—Usability	FAU	Very clear to operate	Very confusing to operate
Most frequent adaptive action—Responsiveness	FAR	Very fast and reliable	Very slow and unreliable

4. Results

4.1. Thermal Comfort Measurement

Table 3 and Figure 4 show the statistical summary of measured environmental parameters in Phase 1 and 2. In both phases, the mean of air temperature was lowest in Room 2 (Phase 1: Mean Ta = 27.95, SD = 0.38 and Phase 2: Mean Ta = 28.68, SD = 1.05). Air temperature in Rooms 2 and 3 was significantly lower than in other rooms. As seen in Figure 4, the first quartile (Q1) for air temperature in Rooms 1, 4, 5 and 6 was above what ASHRAE standard 55 recommended for thermal comfort. In both phases, relative humidity in Rooms 4, 5 and 6 was significantly lower than in other rooms, and the highest was reported in Room 2.

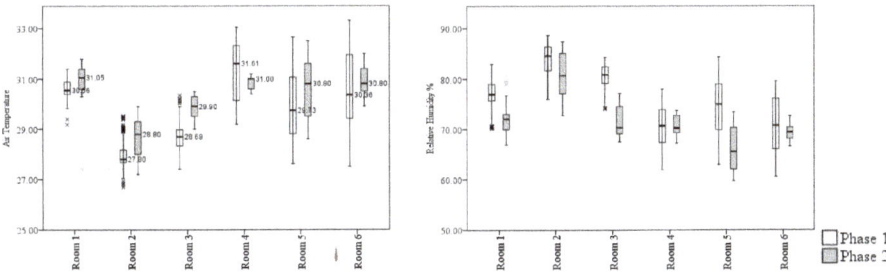

Figure 4. Statistical summary of measured air temperature (°C) and relative humidity (%).

Table 3. Statistical summary of measured thermal parameters.

		Phase 1					Phase 2						
		Mean		SD			Mean			SD			
Participant	Location	Ta (°C)	RH (%)	Ta	RH	Tg (°C)	Ta (°C)	RH (%)	Va (m/s)	Tg	Ta	RH	Va
Subject 1	Room 1	30.60	77.28	0.38	2.69	30.85	30.88	71.32	0.22	1.23	1.05	5.00	0.11
Subject 2	Room 2	27.95	83.97	0.55	2.54	28.92	28.68	80.79	0.38	0.82	0.75	4.07	0.17
Subject 3	Room 3	28.71	80.46	0.57	2.37	30.16	29.82	71.15	0.82	0.43	0.45	2.79	0.11
Subject 4	Room 4	31.34	70.63	1.17	3.77	30.86	30.83	70.65	1.03	0.28	0.24	1.86	0.21
Subject 5	Room 5	29.99	74.20	1.37	5.59	30.41	30.51	66.16	1.02	1.06	1.04	4.27	0.34
Subject 6	Room 6	30.60	70.98	1.40	5.16	31.1	30.97	69.41	1.5	0.56	0.58	1.58	0.18

The ASHRAE adaptive thermal comfort model was used to determine the acceptability of the thermal conditions in the surveyed spaces during measurement of Phase 2. Table 4 reports the calculated operative temperature for each room and the period (percentage) within 80% and 90% acceptability limits specified by the ASHREA adaptive thermal comfort standard. The best measured thermal environment was found in Room 2, which was within an 80% acceptability limit for 93.3% of the surveyed period. The worst measured thermal environment was found in Rooms 1, 4 and 6, within an 80% acceptability limit for less than 5% of the measured period. Rooms 4, 5 and 6 had significantly higher indoor air velocities (Figure 5). The lowest indoor air velocity was reported in Room 1.

Table 4. Indoor operative temperature within American Society of Heating, Refrigerating and Air-Conditioning Engineers (ASHRAE) 80% and 90% acceptability limit.

Room No.	Minimum	Maximum	Mean	Standard Deviation	% Top within 90% Acceptability Limits	% Top within 80% Acceptability Limits
1	24.9	31.7	30.9	1.13	3.4	3.4
2	26.9	30.1	28.9	0.82	54.5	93.3
3	29.2	30.8	30.1	0.43	0	41.7
4	30.4	31.2	30.8	0.28	0	0
5	28.5	32.2	30.4	1.06	9.2	43.8
6	30.2	32.2	31.1	0.56	0	0

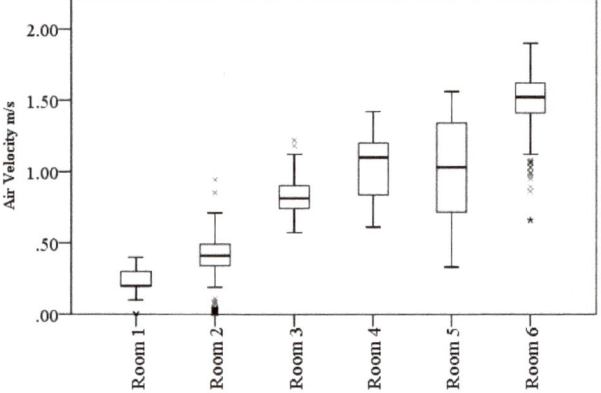

Figure 5. Statistical summary of indoor air velocity in the rooms during Phase 2.

4.2. Thermal Comfort Perception

The mean scores for the occupant's perception of thermal acceptability (TC), thermal sensation vote (TSV) and air movement acceptability (AM) were compared using one-way ANOVA and followed up with post hoc testing to determine which specific subjects evaluated the thermal environment differently. The Levene's test for equality of variances was statistically significant ($p < 0.05$), therefore equal variance was not assumed and results were interpreted using Welch ANOVA and Games–Howell post hoc. The results are summarized in Table 5. In both phases of this survey, the mean scores for TC, TSV and AM were statistically significantly different among the participants.

Table 5. Results of one-way ANOVA for occupant's perception of the thermal environment.

Phase	IEQ Factor	Levene's Test for Equality of Variances		ANOVA			Robust Tests of Equality of Means		
		F	Sig.	F	df (Within Groups)	Sig (Between Groups)	F	df (Within Groups)	Sig. (Between Groups)
Phase 1	TC	7.135	0.000	11.578	5	0.000	17.343	5	0.000
	TSV	10.739	0.000	7.141	5	0.000	8.986	5	0.000
	AM	2.284	0.050	25.368	5	0.000	32.088	5	0.000
Phase 2	TC	5.281	0.001	3.758	5	0.006	5.137	5	0.003
	TSV	5.672	0.000	2.229	5	0.067	11.098	5	0.000
	AM	2.422	0.049	13.369	5	0.000	11.194	5	0.000

TC: thermal acceptability; TSV: thermal sensation vote; AM: air movement acceptability; df: degrees of freedom; Sig.: significance.

Figure 6 shows the means plot for TC, TSV and AM. Games–Howell post hoc analysis revealed the following: in both phases, Subject 2 was the most satisfied of TC (M = 1.67, SD = 1.12); Subjects 1 and 3 were the least satisfied with TC. In Phase 1, Subjects 1 and 3 were significantly more dissatisfied with TC than the others. In Phase 2, Subject 2 was significantly more satisfied with the thermal environment than Subject 1 (1.39, 95% CI [0.3 to 2.47], $p = 0.01$) and Subject 3 (1.67, 95% CI [0.48 to 2.86], $p = 0.004$). During Phase 1, mean scores for TSV reported by Subjects 2 and 3 were significantly lower than those for Subjects 1, 4 and 6. In Phase 2, mean scores for TSV reported by Subject 6 were significantly lower than those for Subject 4 (0.79, 95% CI [0.25 to 1.33], $p = 0.01$) and Subject 1 (0.85, 95% CI [0.14 to 1.56], $p = 0.025$). In both phases, Subject 5 was most satisfied with air movement; while Subjects 1 and 3 were significantly more dissatisfied than the others.

Figure 7 summarizes occupants' preferences for indoor air movements in Phase 1 and 2. The subjects rarely requested to reduce indoor air speeds and mostly preferred an increase in indoor air velocities. In Phase 2, Subject 5 was the most satisfied with indoor air movement and did not request any change. In Phase 2, Subjects 2, 3, 4 and 6 requested an increase in indoor air movement in more than 40% of their responses.

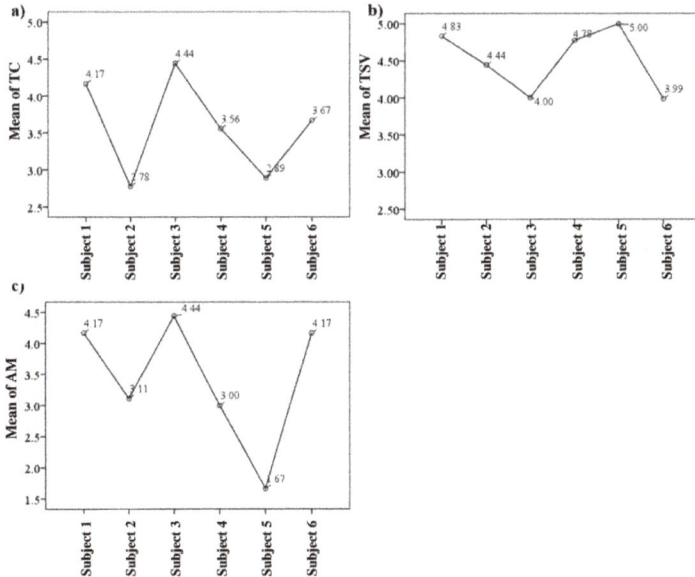

Figure 6. Means plot for (**a**) TC; (**b**) TSV; (**c**) AM.

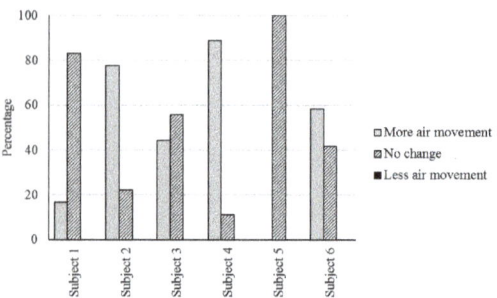

Figure 7. Statistical summary of air movement preference.

4.3. Adaptive Behaviors

Figure 8 summarizes the percentage of various adaptive behaviors performed by each participant during Phase 1 and 2. Consumption of drinks, clothing adjustment, adjustments to windows, blinds, electric fans and doors were among the most preferred adaptive behaviors reported by the subjects. Figure 9 summarizes the frequency of the most intermittent adaptive actions within the last hour prior to filling the survey. Adjustment to the room door (opening or closing) was the most frequently performed behavioral adjustment reported by most of the participants (Subjects 2, 4 and 6). Opening the room door was intended to increase ventilation in the room. Other frequent adaptive behavior includes clothing adjustments reported by Subject 3 and consumption of drinks reported by Subject 5. The participants rated the accessibility (FAA), usability (FAU) and responsiveness (FAR) of their most frequently performed adaptive behavior on a seven-point scale. The mean scores for these three items together with mean of clothing insulation (clo) and metabolic rate (met) were compared using one-way ANOVA and followed up with post hoc testing to determine significant differences between the means. The mean score of metabolic rate was the only statistically non-significant factor among the participants (F [18.67] = 1.73, $p = 0.187$).

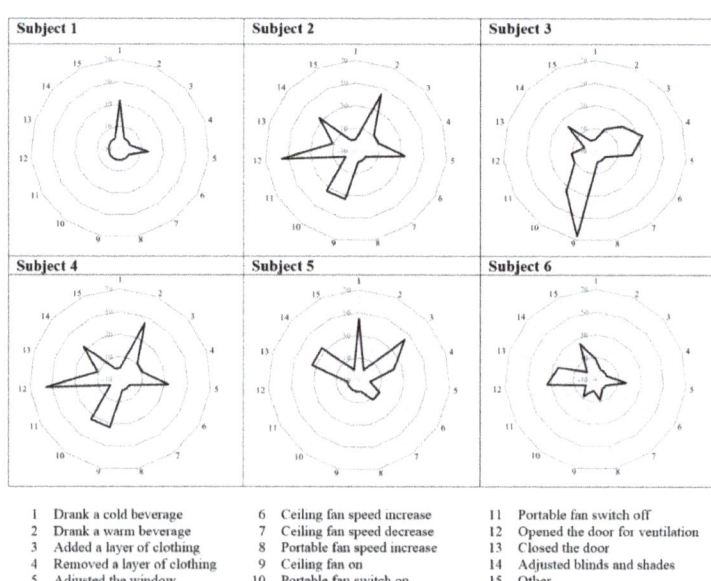

1	Drank a cold beverage	6	Ceiling fan speed increase	11	Portable fan switch off
2	Drank a warm beverage	7	Ceiling fan speed decrease	12	Opened the door for ventilation
3	Added a layer of clothing	8	Portable fan speed increase	13	Closed the door
4	Removed a layer of clothing	9	Ceiling fan on	14	Adjusted blinds and shades
5	Adjusted the window	10	Portable fan switch on	15	Other

Figure 8. Frequency of adaptive behaviors.

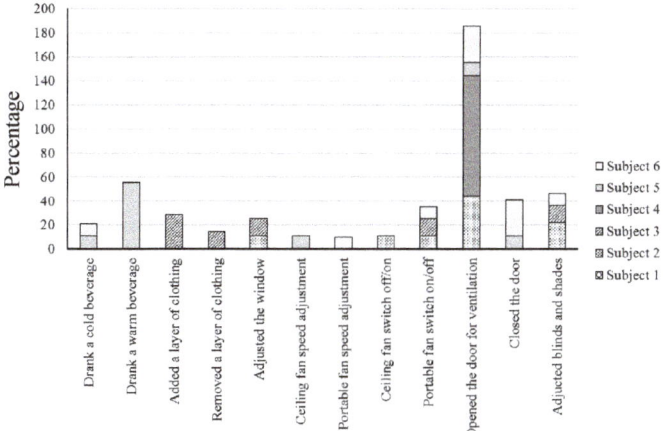

Figure 9. Most frequent adaptive behaviors.

Figure 10 shows the means plot for accessibility, usability, responsiveness and clothing insulation. Subject 1 did not report a response for FAA, FAU and FAR. Subject 5 was most satisfied with accessibility and usability of the most frequent adaptive actions. Subjects 2, 5 and 6 were significantly more satisfied with accessibility and usability than Subject 4, who was the least satisfied for both. The highest mean difference for accessibility is between Subject 4 and 5 (2.44, 95% CI [1.37 to 3.52], $p = 0.0005$) and the lowest mean difference is between Subject 4 and 6 (0.78, 95% CI [0.70 to 2.86], $p = 0.0005$). The highest mean difference for usability is between Subject 4 and 5 (2.0, 95% CI [1.39 to 2.61], $p = 0.0005$) and lowest mean difference is between Subject 4 and 2 (1.11, 95% CI [0.49 to 1.74], $p = 0.006$). Subject 2 was significantly more satisfied with the responsiveness of her most frequent adaptive actions than the others. The highest mean difference for responsiveness is between Subject 6 and 2 (1.72, 95% CI [0.66 to 2.79], $p = 0.001$). The highest clothing insulation was reported for Subject 3 (M = 0.45, SD = 0.04). The highest mean difference for clothing insulation is between Subject 3 and 5 (0.21, 95% CI [0.16 to 0.25], $p = 0.0005$).

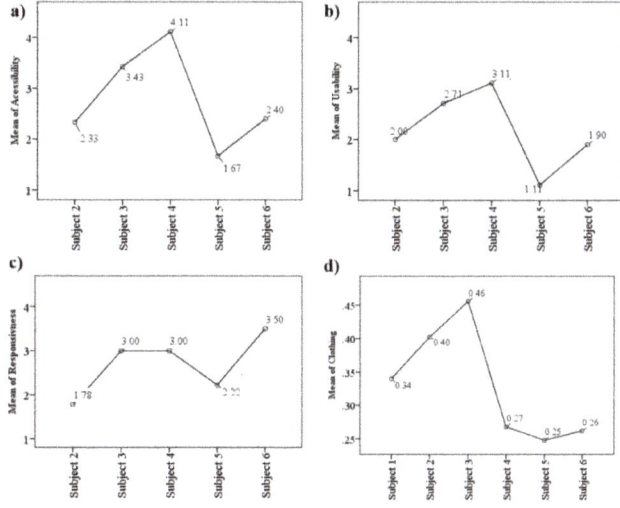

Figure 10. Means plot for (**a**) accessibility; (**b**) usability; (**c**) responsiveness and (**d**) clothing insulation.

4.4. Environmental Controls

Figure 11 summarizes the state of environmental controls such as windows, blinds and doors in the surveyed rooms. The window position was consistent throughout the survey for all subjects except Subject 6. The window was kept either partially opened or fully opened by all participants. The participants mainly interacted with the room door and blinds. Subject 1, who had the lowest indoor air velocity, also had the room door closed throughout the survey. Subject 4 and 6 in Building-B had window blinds fully closed for 27–50% of the survey duration. In Building B, the adjustable air vent on the wall allows ventilation from the building envelope even when the window blinds are closed. This may have allowed Subjects 4 and 6 to maintain cross ventilation in the room when the windows blinds were fully closed. Those in Building B also used electric lights for over 50% of the surrey duration.

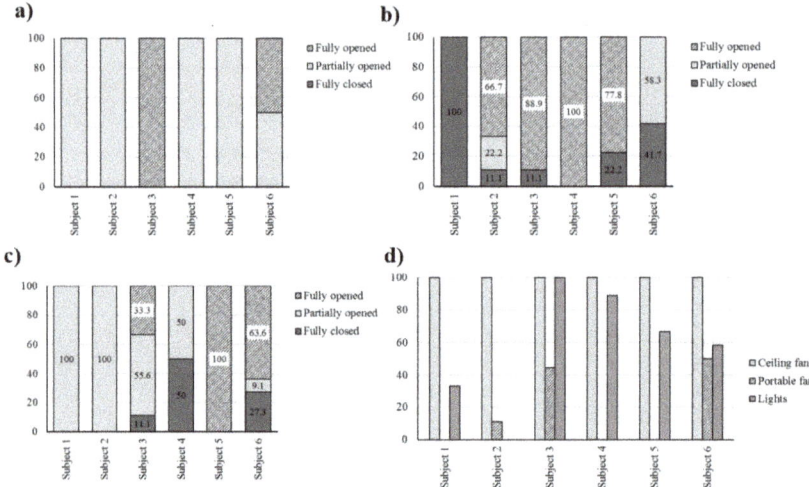

Figure 11. State of environmental controls in the surveyed rooms (**a**) position of windows; (**b**) position of doors; (**c**) position of blinds; (**d**) controls operating during survey.

4.5. Indoor Environment Quality

The state of the thermal environment indoors may significantly influence an occupant's perception of other indoor environment quality factors. Further behavioral adjustments aimed at thermal comfort alleviation in naturally ventilated buildings may also result in modifications to these indoor environment quality factors such as air quality, lighting and acoustics. In this survey, eleven questionnaire items were used to evaluate the occupant's perception towards air quality, lighting and acoustics. Spearman's rank-order correlation was used to identify significant correlations between occupants' perceptions of measured indoor environment quality variables.

Table 6 summarizes the results of Spearman's rank-order correlation for occupant's perception of various indoor environment quality factors. Occupant's acceptance of the thermal environment was significantly and positivity correlated with air movement acceptance (Spearman's correlation coefficient $\rho = 0.558$, $n = 54$, $p = 0.0005$), air quality acceptance ($\rho = 0.468$, $n = 54$, $p = 0.0005$) and air freshness ($\rho = 0.649$, $n = 54$, $p = 0.0005$). Further occupants associated higher draft with greater acceptability of the air quality ($\rho = 0.391$, $n = 54$, $p = 0.003$) and air movements ($\rho = 613$, $n = 54$, $p = 0.0005$), which indicated that higher indoor air velocities might be preferred by the participants and higher air velocities might also increase occupant's satisfaction towards the indoor air quality. Spearman's correlation indicates that acceptance of lighting quality was significantly and positively correlated to amount of glare from to daylight ($\rho = 0.501$, $n = 53$, $p = 0.0005$) and electric lights ($\rho = 0.491$, $n = 53$, $p = 0.0005$).

Further analysis was conducted to identify significant differences in mean score for occupant perception of glare from daylight (NLG), glare from electric lights (AGL) and noise. The mean scores for NGL, AGL and noise were compared using one-way ANOVA and followed up with post hoc testing to determine significant differences between the means. The results are summarized in Table 7. Mean scores for NGL, AGL and noise were statistically significantly different. Figure 12 shows the means plot for NGL, AGL and noise. Subjects 4 and 6 who had the window blind closed during part of the survey, reported significantly less glare from day lighting. The highest mean difference for NGL is between Subjects 3 and 4 (3.0, 95% CI [1.67 to 4.33], $p = 0.0005$) and lowest mean difference is between Subjects 1 and 6 (1.5, 95% CI [0.09 to 2.91], $p = 0.03$). The highest mean difference for AGL is between Subjects 3 and 6 (3.69, 95% CI [2.85 to 4.54], $p = 0.0005$) and lowest mean difference is between Subjects 2 and 5 (1.0, 95% CI [0.1 to 1.90], $p = 0.022$). Subject 2 was significantly more satisfied with indoor noise than all others. The highest mean difference for noise is between Subjects 6 and 2 (2.78, 95% CI [1.38 to 4.17], $p = 0.0005$).

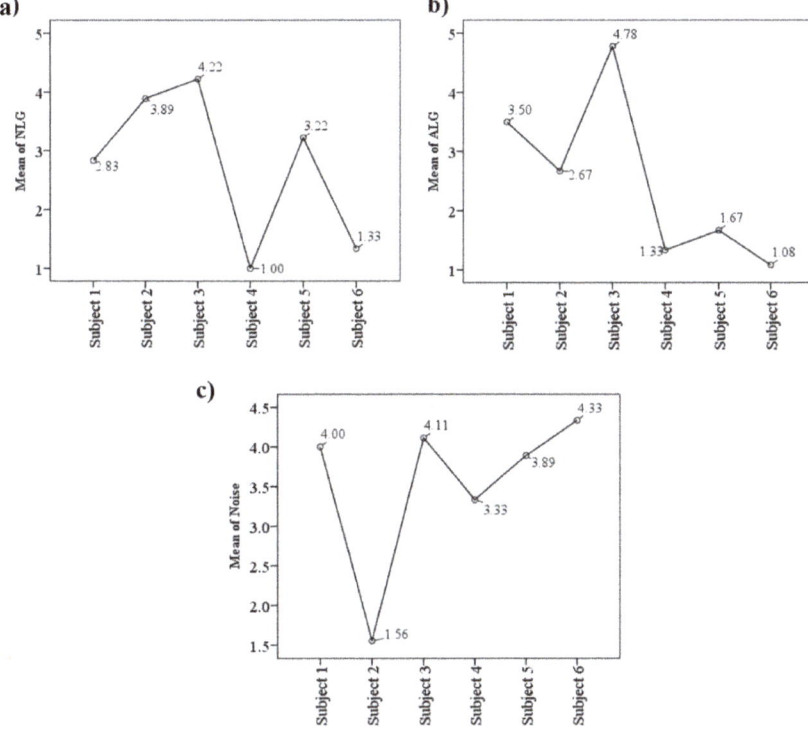

Figure 12. Means of (**a**) glare from daylight; (**b**) glare from electric lights; (**c**) noise.

Table 6. Results of spearman's rank-order correlation for the perception of indoor environment quality.

	TSV	AMA	AQA	DRAFT	HUMID	AF	LQA	NLA	NLG	ALA	ALG	NOISE
TC	0.351 **	0.558 **	0.486 **	−0.157	0.403 **	0.649 **	0.193	−0.070	0.075	−0.052	0.234	0.149
TSV		−0.124	0.006	0.189	0.248	0.121	0.253	−0.032	0.213	−0.016	0.136	−0.065
AMA			0.550 **	−0.613 **	0.298 *	0.438 **	0.173	−0.108	0.036	−0.100	0.182	0.278 *
AQA				−0.391 **	0.241	0.522 **	0.287 *	0.061	0.173	0.045	0.370 **	0.189
DRAFT					0.036	−0.109	−0.143	0.072	−0.200	0.298 *	−0.205	0.087
HUMID						0.028	−0.024	−0.331 *	−0.158	0.170	0.151	0.007
AF							0.282 *	0.304 *	0.241	0.162	0.249	0.311 *
LQA								0.390 **	0.501 **	0.327 *	0.491 **	0.179
NLA									0.578 **	0.247	0.382 **	0.123
NLG										0.077	0.650 **	−0.133
ALA											0.283 *	0.271 *
ALG												−0.172

** Correlation is significant at the 0.01 level (2-tailed). * Correlation is significant at the 0.05 level (2-tailed).

Table 7. Results of one-way ANOVA for NGL, AGL and noise.

IEQ Factor	Levene's Test for Equality of Variances		ANOVA			Robust Tests of Equality of Means		
	F	Sig.	df (Within Groups)	F	Sig. (Between Groups)	df (Within Groups)	F	Sig. (Between Groups)
NGL	1.92	0.108	48	17.44	0.000	19.92	21.14	0.000
AGL	3.25	0.013	48	4.48	0.000	18.22	38.375	0.000
Noise	5.04	0.001	48	8.54	0.000	18.49	6.989	0.001

5. Summary of Findings

Firstly, occupants in Building B were more likely to accept the thermal environments which were not within those specified by the ASHRAE thermal comfort standard for naturally conditioned spaces. A significant influence came from indoor air velocity, which was linked to the occupant's acceptance of these imposing heat stresses. Those who were either exposed to higher air velocities (Subject 6) or those who were significantly more satisfied with indoor air moment (Subjects 4 and 5) showed greater tolerance to the heat stress.

Secondly, all occupants made a considerable attempt to increase air velocity in their rooms. The ceiling fans in all the surveyed rooms were switched on during the whole survey period. Some occupants also used a portable fan (Subjects 2, 3 and 6). The occupants regarded cross ventilation as an effective mechanism to increase indoor air velocity. All surveyed rooms could only be cross-ventilated by opening both windows and room doors. The occupants frequently changed the position of the door (open/closed) while the window was kept in a constant position, either fully opened or partially opened. However, opening the door for cross ventilation was regarded as highly responsive for heat alleviation only when the thermal environment was within the acceptable condition specified by the ASHRAE comfort standard for naturally conditioned spaces.

Thirdly, personal adjustments such as lower clothing insulation and consumption of drinks were linked to a significant improvement in their perceptions of the thermal environment. Significantly lower clothing insulation value was reported by those who were more likely to accept a less than preferred thermal environment (Subjects 4, 5 and 6). Furthermore, those who performed personal adjustments such as consumption of drinks were also significantly more satisfied with the accessibility, usability and responsiveness of this action.

Last but not least, the occupants' perception of the thermal environments was significantly associated with other indoor environmental factors such as indoor air quality. Higher satisfaction with the thermal environment and air movement acceptance were associated with greater satisfaction of air quality and air freshness. However, the cause of this association cannot be determined. High association between occupants' acceptance of air movement, higher draft and greater acceptance of air quality indicated the positive contribution of higher air velocities in naturally ventilated buildings to increase occupants' perception of air quality. This finding adds to the positive attribute of higher indoor air velocity to improve psychological cooling effect of occupants in naturally ventilated buildings.

In summary, adaptive behaviors such as window and door opening, changing clothing conditions and consumptions of cold drinks have an important role in helping occupants accept thermal environments beyond the comfort zone. The window opening behavior was also found effective in temperate climates such as Denmark where residents opened windows according to outdoor climates [28]. Research in other climates showed that these adaptive behaviors in naturally ventilated or mixed-mode conditions were closely linked to and can be predicted by outdoor temperature [29,36]. These precedent studies support the present study that adaptive behaviors can help occupants in naturally ventilated buildings in tropical climates adapt to thermal environments beyond the comfort zone.

6. Conclusions

This study revealed that occupants living in naturally ventilated buildings in tropics were exposed to higher operative temperatures than the ASHRAE comfort standard for naturally conditioned spaces. However, they were more likely to accept such conditions. Two behavioral adjustments were found to have a profound impact on occupants' acceptance of the imposing heat stresses: (1) increasing the indoor air velocity by turning on mechanical fans and opening the door/windows for cross ventilation and (2) lowering clothing insulation values by changing clothes and dressing in fewer clothes. These behavioral adjustments successfully alleviated thermal stress as they directly impacted the body's heat balance. The efficacy of available control as a form of adaption [37,38] beyond the body's heat balance seemed to have little impact on improving the occupants' acceptance of the thermal environment. In tropical climates, incorporating supplementary ventilation mechanisms such as adjustable air vents on the building envelope could positively contribute towards occupants' satisfaction of the indoor environment, as occupants could lower window blinds to let outdoor air in while avoiding discomfort glare. Supplementary mechanical ventilation, such as fans, is instrumental in increasing indoor air velocities and improving thermal comfort. Higher air velocity also results in greater satisfaction of indoor air quality.

This study is a pilot study. It has two limitations: (1) the sample size is small; and (2) the outdoor climate is missing in this study. Future studies should enlarge the sample size. It should also correlate indoor and outdoor climates. The gender and ethnical backgrounds should be considered for an understanding of their comfort expectations. A future study will use sophisticated statistical methods to estimate the probability of a certain adaptive behavior happening as a function of temperature variations. The common statistical method used in similar studies is multiple logistic regression technique [30–32,36]. Usually, the adaptive behaviors should be analyzed separately: using each behavior as the dependent variable and the outdoor air temperature as the dependent variable [28]. The logistic regression models can help to predict the probability of residents using a particular strategy to achieve comfort, as a function of the outdoor air temperature recorded from the nearest weather station [29].

Acknowledgments: The authors are grateful to all participants for their continuous inputs during this longitudinal survey. The authors thank the Department of Architecture, National University of Singapore.

Author Contributions: All authors contributed equally to this paper.

Conflicts of Interest: The authors declare no conflict of interest.

References

1. Brandon, P.S.; Lombardi, P. *Evaluating Sustainable Development in the Built Environment*; John Wiley & Sons: Hoboken, NJ, USA, 2010.
2. Pérez-Lombard, L.; Ortiz, J.; Pout, C. A review on buildings energy consumption information. *Energy Build.* **2008**, *40*, 394–398. [CrossRef]
3. Gou, Z.; Lau, S.-Y.S.; Lin, P. Understanding domestic air-conditioning use behaviours: Disciplined body and frugal life. *Habitat Int.* **2017**, *60*, 50–57. [CrossRef]
4. Gamage, W.; Lau, S.; Qin, H.; Gou, Z. Effectiveness of air-well type courtyards on moderating thermal environments in tropical Chinese Shophouse. *Archit. Sci. Rev.* **2017**, *60*, 493–506. [CrossRef]
5. Building and Construction Authority. BCA Building Energy Benchmarking Report 2015. Available online: https://www.bca.gov.sg/GreenMark/others/BCA_BEBR_Abridged_FA_2015.pdf (accessed on 29 June 2016).
6. Building and Construction Authority. Building Planning and Massing. Green Building Platinum Series. Available online: https://www.bca.gov.sg/GreenMark/others/bldgplanningmassing.pdf (accessed on 30 July 2016).
7. American Society of Heating, Refrigerating and Air Conditioning Engineers (Atlanta, Georgia). *ANSI/ASHRAE Standard 55-2013: Thermal Environmental Conditions for Human Occupancy*; ASHRAE: New York, NY, USA, 2013.
8. Meteorological Service Singapore. *Records of Climate Station Mean: (Climatological Reference Period: 1984–2015)*; Meteorological Service Singapore: Singapore, 2016.
9. American Society of Heating, Refrigerating and Air Conditioning Engineers (Atlanta, Georgia). *ANSI/ASHRAE Standar 55-1992: Thermal Environmental Conditions for Human Occupancy*; ASHRAE: New York, NY, USA, 1992.
10. Milne, M. Climate Consultant v6.0. Available online: http://www.energy-design-tools.aud.ucla.edu/climate-consultant/request-climate-consultant.php (accessed on 30 June 2016).
11. Liping, W.; Hien, W.N. Applying natural ventilation for thermal comfort in residential buildings in Singapore. *Archit. Sci. Rev.* **2007**, *50*, 224–233.
12. Ang, B.; Goh, T.; Liu, X. Residential electricity demand in Singapore. *Energy* **1992**, *17*, 37–46. [CrossRef]

13. Chuan, L.; Ukil, A. Modeling and validation of electrical load profiling in residential buildings in Singapore. *IEEE Trans. Power Syst.* **2015**, *30*, 2800–2809. [CrossRef]

14. National University of Singapore. *NUS Sustainability Strategic Plan 2017–2020*; Office of Environmental Sustainability, National University of Singapore: Singapore, 2017.

15. Nicol, J.F.; Humphreys, M.A. Adaptive thermal comfort and sustainable thermal standards for buildings. *Energy Build.* **2002**, *34*, 563–572.

16. Nicol, F. Adaptive thermal comfort standards in the hot-humid tropics. *Energy Build.* **2004**, *36*, 628–637. [CrossRef]

17. Mallick, F.H. Thermal comfort and building design in the tropical climates. *Energy Build.* **1996**, *23*, 161–167. [CrossRef]

18. Dutt, A.; De Dear, R.; Krishnan, P. Full scale and model investigation of natural ventilation and thermal comfort in a building. *J. Wind Eng. Ind. Aerodyn.* **1992**, *44*, 2599–2609. [CrossRef]

19. Baker, N.; Standeven, M. Thermal comfort for free-running buildings. *Energy Build.* **1996**, *23*, 175–182. [CrossRef]

20. De Dear, R.J.; Brager, G.S.; Reardon, J.; Nicol, F. Developing an adaptive model of thermal comfort and preference/Discussion. *ASHRAE Trans.* **1998**, *104*, 145.

21. Gou, Z.; Lau, S.S.-Y.; Chen, F. Subjective and Objective Evaluation of the Thermal Environment in a Three-Star Green Office Building in China. *Indoor Built Environ.* **2012**, *21*, 412–422. [CrossRef]

22. Feriadi, H. Thermal Comfort for Naturally Ventilated Residential Buildings in Tropical Climate. Ph.D. Thesis, National University of Singapore, Singapore, 2004.

23. Wong, N.H.; Khoo, S.S. Thermal comfort in classrooms in the tropics. *Energy Build.* **2003**, *35*, 337–351. [CrossRef]

24. Brager, G.S.; de Dear, R.J. Thermal adaptation in the built environment: a literature review. *Energy Build.* **1998**, *27*, 83–96. [CrossRef]

25. Wohlwill, J.F. Human adaptation to levels of environmental stimulation. *Hum. Ecol.* **1974**, *2*, 127–147. [CrossRef]

26. Fountain, M.; Brager, G.; de Dear, R. Expectations of indoor climate control. *Energy Build.* **1996**, *24*, 179–182. [CrossRef]

27. Fanger, P.O. *Thermal Comfort: Analysis and Applications in Environmental Engineering*; Danish Technical Press: Copenhagen, Denmark, 1970.

28. Andersen, R.V.; Toftum, J.; Andersen, K.K.; Olesen, B.W. Survey of occupant behaviour and control of indoor environment in Danish dwellings. *Energy Build.* **2009**, *41*, 11–16. [CrossRef]

29. Kim, J.; de Dear, R.; Parkinson, T.; Candido, C. Understanding patterns of adaptive comfort behaviour in the Sydney mixed-mode residential context. *Energy Build.* **2017**, *141*, 274–283. [CrossRef]

30. Haldi, F.; Robinson, D. On the behaviour and adaptation of office occupants. *Build. Environ.* **2008**, *43*, 2163–2177. [CrossRef]

31. Nicol, J.F.; Humphreys, M.A. A Stochastic Approach to Thermal Comfort-Occupant Behavior and Energy Use in Buildings. *ASHRAE Trans.* **2014**, *110*, 554–568.

32. Rijal, H.B.; Tuohy, P.; Humphreys, M.A.; Nicol, J.F.; Samuel, A.; Clarke, J. Using results from field surveys to predict the effect of open windows on thermal comfort and energy use in buildings. *Energy Build.* **2007**, *39*, 823–836. [CrossRef]

33. Ackerly, K.; Brager, G.; Arens, E. *Data Collection Methods for Assessing Adaptive Comfort in Mixed-Mode Buildings and Personal Comfort Systems*; Center for the Built Environment, University of California: Oakland, CA, USA, 2012.

34. Gou, Z.; Prasad, D.; Lau, S.S.-Y. Are green buildings more satisfactory and comfortable? *Habitat Int.* **2013**, *39*, 156–161. [CrossRef]

35. Gou, Z.; Lau, S.S.-Y.; Shen, J. Indoor Environmental Satisfaction in Two LEED Offices and its Implications in Green Interior Design. *Indoor Built Environ.* **2012**, *21*, 503–514. [CrossRef]

36. Schweiker, M.; Shukuya, M. Comparison of theoretical and statistical models of air-conditioning-unit usage behaviour in a residential setting under Japanese climatic conditions. *Build. Environ.* **2009**, *44*, 2137–2149. [CrossRef]

37. Frontczak, M.; Wargocki, P. Literature survey on how different factors influence human comfort in indoor environments. *Build. Environ.* **2011**, *46*, 922–937. [CrossRef]
38. Williams, R. Field investigation of thermal comfort, environmental satisfaction and perceived control levels in UK office buildings. In Proceedings of the Healthy Buildings, Milan, Italy, 10–14 September 2015.

 © 2018 by the authors. Licensee MDPI, Basel, Switzerland. This article is an open access article distributed under the terms and conditions of the Creative Commons Attribution (CC BY) license (http://creativecommons.org/licenses/by/4.0/).

Article

Characteristics of Thermal Comfort Conditions in Cold Rural Areas of China: A Case study of Stone Dwellings in a Tibetan Village

Bin Cheng [1], Yangliu Fu [1],*, Maryam Khoshbakht [2], Libin Duan [1], Jian Zhang [2] and Sara Rashidian [3]

[1] College of Civil Engineering and Architecture, Southwest University of Science and Technology, Mianyang 621010, Sichuan, China; chengbin@swust.edu.cn (B.C.); duanlibin@mails.swust.edu.cn (L.D.)

[2] School of Engineering and Built Environment, Griffith University, Gold Coast, QLD 4215, Australia; m.kh@griffithuni.edu.au (M.K.); jian.zhang@griffithuni.edu.au (J.Z.)

[3] Faculty of Society and Design, Mirvac School of Sustainable Development, Bond University, Robina, QLD 4226, Australia; srashidi@bond.edu.au

* Correspondence: fuyangliu@mails.swust.edu.cn; Tel.: +86-187-840-75436

Received: 2 March 2018; Accepted: 20 March 2018; Published: 26 March 2018

Abstract: This paper focuses on thermal environmental conditions in the stone dwellings of a Tibetan village in Danba County, Sichuan, China, in winter. During the study, field measurements and subjective survey studies were collected, simultaneously, to provide a comprehensive understanding of the thermal comfort conditions that were experienced by residents in cold rural areas of Sichuan. Subjective surveys involved questions about thermal comfort perceptions and acceptability in cold conditions. The status of thermal comfort and characteristics of indoor environmental qualities were investigated in the study. The majority of survey participants (47% and 74%) voted as "slightly cool" for temperature, and "slightly dry" for humidity in the studied typical winter days, respectively. The available adaptive opportunities for the residents were investigated through the survey studies. Adjusting clothing, drinking hot beverages, blocking air infiltration through windows, and changing activities were the most common adaptive measures. An adaptive coefficient (λ) was determined based on adaptive predicted mean votes (aPMV) models using least square methods to assess the different adaptation measures in the region. Findings of this study provided a valuable reference for thermal comfort adaptations in cold climates, where limited adaptive opportunities were available due to the low standard of living.

Keywords: Tibetan village in Sichuan Province; thermal environment; adaptive thermal comfort; adaptive coefficient; stone dwelling

1. Introduction

Thermal comfort is an important aspect of residential buildings because it strongly influences occupant health and wellbeing [1]. In the past, research has focused on non-residential buildings [2–7], and occupant thermal perceptions and comfort in residential buildings have been largely overlooked. In thermal comfort studies, two methods were employed: chamber studies, and real building experiments [8,9]. Based on chamber experiments, Fanger developed a heat balance model of thermal comfort for air-conditioned buildings, which estimates predicted mean votes (PMV) and the percentage of predicted dissatisfaction (PPD) [10]. Fanger was one of the first who studied parameter affecting indoor environmental qualities [11]. The PMV model of thermal comfort established the theoretical basis for thermal comfort studies of interactions between a human body and physical parameters. The PMV model became the benchmark for the International Organization for Standardization (ISO) for thermal comfort [12–15]. The PMV model was originally developed based on air conditioning buildings. Several researchers attempted to verify the applicability of the PMV model in naturally ventilated buildings [16–18]. Field measurements, as one assessment tool for post occupancy evaluations [19], indicated that a large deviation exists in the application of the PMV model in the early stage of indoor climate.

Later, De Dear and Brager [20] collected large sample data from a series of field studies in naturally ventilated buildings in the global climate region, and established an adaptive thermal comfort model. The adaptive thermal comfort model showed that occupants exhibit a wider acceptable temperature range in naturally ventilated buildings and in the presence of adaptive measures. Adaptive opportunities, such as changing clothing or air movement helped occupants to make physiological and psychological adjustments [21].

However, research has shown that cultural, social, and climate factors have a strong influence on human thermal comfort [22]. Human bodies respond to thermal environments by physiological adaptations through thermoregulatory systems and create a heat balance [8]. Behavioural adaptations are performed when humans feel thermally dissatisfied with their environment. As shown by De Dear and Brager [23], psychologically adaptive behaviours plays a significant role in determining thermal sensation votes [18,24]. Based on this discussion, occupant psychological and behavioural adaptations in different environmental conditions and their influence on thermal comfort is essential. Yao et al. [8] proposed a theoretical aPMV that considered psychological, physiological, and behavioural adaptability and other influencing factors on thermal comfort sensations. As literature suggests, the aPMV seems to be the most appropriate model for the assessment of cultural, social and climatic impacts on thermal comfort status in rural non-air-conditioned residential buildings [25].

With its diverse culture and climate, China has attracted the attention of many researchers working to develop thermal comfort models with respect to climate, and social and cultural differences in cities such as Guangzhou, Zhejiang, Shanghai, Changsha, Chongqing, Xi'an, Beijing and Harbin [26–35]. Chinese rural residential buildings have also attracted the attention of many scholars who have examined thermal comfort status in various climatic zones in China. Zhou [36] investigated the thermal comfort of rural houses in western Hunan with its hot summer and cold winters, and estimated the adaptive coefficient (λ) of 0.49 for the region. Wang [37] adopted the aPMV model and predicted the average thermal sensation of primary and middle school classrooms in Qinghai rural areas with their severe cold winter seasons. Li [38] studied the Chongqing rural area with its hot summers and cold winters, and simulated the indoor thermal environment. Yang [39] employed the aPMV model to evaluate the indoor thermal environment in timber residential buildings in the Chongqing mountains, and showed that timber houses possess a strong climate adaptability. Zheng [40] conducted field thermal comfort studies in Xi'an's cold climate, and examined the thermal comfort status in four different seasons in the region.

Some researchers have worked on thermal comfort characteristics in the harsh climate of high altitude areas. Wang [41] studied the indoor temperature and humidity conditions of indoor environments in houses at high altitude in the Tibet region, and estimated the comfortable temperature range and the adaptive coefficient in the area. He Q [42] conducted research on the assessment of indoor thermal conditions in Tibetan residences on the West Sichuan Plateau, and suggested a number of design strategies to improve indoor thermal environmental conditions for the residents. Ouyang [43] examined Tibetan dwellings on the Western Sichuan Plateau during winter, and recommended the use of solar energy resources to improve indoor environmental qualities for the residents. Chen [44] measured indoor temperature and humidity levels in Tibetan Danba in transitional seasons, and compared the indoor environmental conditions between spring and summer in the stone houses.

This research aims to show the adaptive level of the human body in the naturally ventilated stone dwellings in the cold climate zone of Sichuan, China, in a winter season. The objective of the research is to examine the effect of the natural environment on occupant thermal comfort status and characteristics. The important issues of housing environment and the needs of residents, particularly in low-income groups, have been raised by many researchers [45]. Due to the harsh climate, and impoverished villagers, a comfortable and healthy indoor environment is difficult to guarantee. Therefore, the study of the thermal environment inside stone dwellings of a Tibetan village in the winter season provides an opportunity to study the lowest acceptable temperature in buildings that have limited heating systems available for residents. This paper offers an opportunity to understand occupant physiological and psychological adaptability to cold climates.

2. Background Information

2.1. Climate Characteristics

The Tibetan Plateau is known as "the roof of the world" with an average altitude of 4950 m above the sea, and its exposure to unobstructed freezing arctic air from the north. The air is severely dry for the majority of the year, with an average annual snowfall of only 46 cm. The studied buildings are located in a Tibetan village in Danba County of Sichuan Province, China. The location of Danba County is illustrated in Figure 1. The climate is characterised by a strong variation in the length of day during different seasons of the year.

Figure 1. Map of location of the studied area.

Based on the Köppen climate classification, Sichuan Province is divided into five classes, from temperate to severely cold, all with different climate characteristics. However, cold rural areas of Sichuan include the central region of Garze, the southern part of Aba, and the northwest edge of

Mianyang and Yaan (see Figure 2). Due to the high altitude, Tibetan villages in the cold rural area of Sichuan have higher solar radiation and lower annual average air temperature compared to the other cities in the same building climate zone. Table 1 shows the meteorological parameter of a typical county in the cold rural area of Sichuan. The average annual air temperature of these typical cities in the country is less than 15 °C. In the hottest month, the highest average temperature is below 25 °C. The average sunshine duration is more than 1800 h per year, which shows rich solar energy availability in these areas.

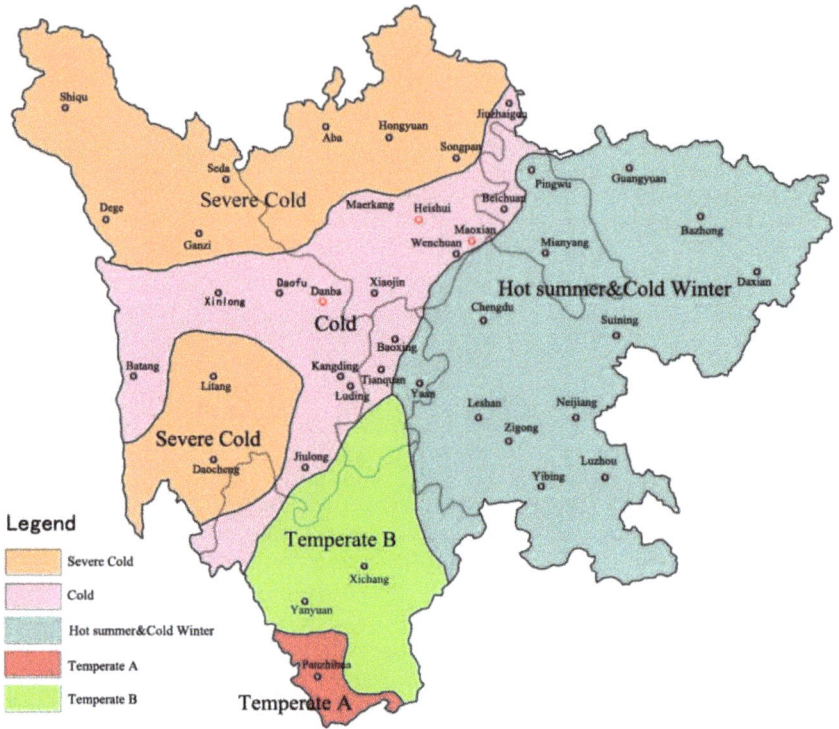

Figure 2. Building climate demarcation of Sichuan province.

Table 1. Comparison of the meteorological parameters of typical county in cold rural area of Sichuan.

County	North Latitude [1]	East Longitude [1]	Annual Average Air Temperature (°C)	Average Temperature of the Coldest Month (°C)	Average Temperature of the Hottest Month (°C)	Annual Average Sunshine Duration (h)	Annual Altitude (m)
Danba County	30°52′	101°53′	14.2	4.4	22.4	2106.9	1800
Heishui County	32°03′	102°59′	9	5.4	10.9	2417	2350
Mao County	31°41′	103°51′	11.1	0.8	20.4	1549.4	1580

[1] Observatory site.

2.2. Buildings

The studied stone dwellings are typically three to five storeys high. External walls are constructed from shale stone or rammed earth, with load-bearing walls measuring 60 cm in thickness. Partition walls are also composed of shale Stone. Floors are covered by timber boards, while roofs are composed of stone and clay. The timber framed windows are single-glazed with an orientation towards the south. No particular patterns in the orientation of the buildings were identified. However, to maximise solar gains, buildings are built on the sunny side of the mountain. No central heating systems are available for residents to keep their houses warm in winter. Typical Tibetan stone dwellings in Danba County are presented in Figure 3.

Figure 3. Typical Tibetan stone dwellings in Danba County.

3. Methodology

The methodology for this study consisted of both subjective and objective evaluations. The subjective study was conducted by administering questionnaires to occupants of the studied buildings. The objective evaluations were performed by measuring indoor and outdoor environmental qualities. The studied period represents the typical winter climate in Danba County. Both subjective and objective evaluations were performed every 1-h from 8:00 a.m. and 9:00 p.m. for the period from 5 January through 10 January in 2018. Statistical techniques were used for data analysis.

3.1. Subjective Questionnaire

Subjective evaluations are one of the most important aspects of thermal comfort studies. The survey of this study aimed to investigate thermal comfort sensations, acceptability, and thermal preferences. The questionnaires were divided into two sections concerning: general demographic information, and occupant comfort sensation. The first section asked about participant basic demographic information: namely, age, gender, town of origin, clothing, and activity. ASHRAE Standard 55-1981 was used to calculate occupant clothing insulation (clo) and metabolic activity (met) [46]. The questions about comfort in the second section of the questionnaire asked participants about their real-time sensation regarding (1) temperature, (2) humidity, and (3) air velocity.

The thermal comfort sensation (TSV) question was designed based on the recommendation by the ASHRAE seven-point scale of thermal sensations (−3 cold, −2 cool, −1 slightly cool, 0 neutral, +1 slightly warm, +2 warm, +3 hot) [47]. Similarly, the other two sensation questions about humidity and air velocity were design based on ASHRAE 7-point scale answers. For humidity, participants were asked to vote as 0 for neutral feeling, +3 for very humid sensations, and −3 for very dry sensations. For air velocity, participants were asked to express their sensation about air movement by scoring 0 for neutral feeling, +3 for very still, and −3 for very breezy. More details of the scale point divisions for the three comfort sensation questions are presented in Table 2.

Table 2. The scale of thermal, humidity and air velocity.

Scale Points	Comfort Sensation		
	Temperature	Humidity	Air Velocity
+3	Hot	Very humid	Very still
+2	Warm	Humid	Still
+1	Slightly warm	Slightly humid	Slightly still
0	Neutral	Neutral	Neutral
−1	Slightly cool	Slightly dry	Slightly breezy
−2	Cool	Dry	Breezy
−3	Cold	Very dry	Very breezy

3.2. Environmental Parameter Measurements

Indoor thermal environment field surveys in this study were conducted in the stone dwellings of Danba County in a typical week in winter (from 01/05 to 01/10). The indoor thermal environment was analysed by measuring indoor and outdoor air temperature, globe temperature, humidity, and air velocity. Some information about the measuring equipment, valid range, and accuracy of the devices is summarized in Table 3. Research has shown that the accuracy of measurement tools can significantly influence thermal environment assessments [48]. The recommended accuracy levels by ISO 7726 are included in Table 3. The accuracy of the tools is according to the recommendations by ISO 7726 [28]. Figure 4 illustrates the picture of the instruments which were used in the measurement campaign. In total three measurement instruments were used for measuring air temperature, globe temperature, air velocity, and relative humidity. Air temperatures were measured by an air thermometer using JTR05. Globe temperatures and air velocity were measured using JTSOFI-IAQ with an embedded Globe thermometer and an anemometer. Relative humidity levels were measured with a hygrometer using a Micro Log.

Table 3. Monitoring parameters and specification of equipment.

Monitoring Parameters	Instrument/Sensor	Trade Name	Valid Range	Accuracy	Required Accuracy by Standards
Air temperature	Air Thermometer	JTR05	−20~+120 °C	±0.3 °C	± 0.5 °C [14]
Globe temperature	Globe thermometer	JTSOFI-IAQ	−20~+50 °C	±0.3 °C	±0.4 °C [49]
Air velocity	Anemometer	JTSOFI-IAQ	0.05~2 m/s	±0.03 m/s	±0.05 m/s [14]
Relative humidity (RH)	Hygrometer	Micro Log	0~100%	±2%	±3% [49]

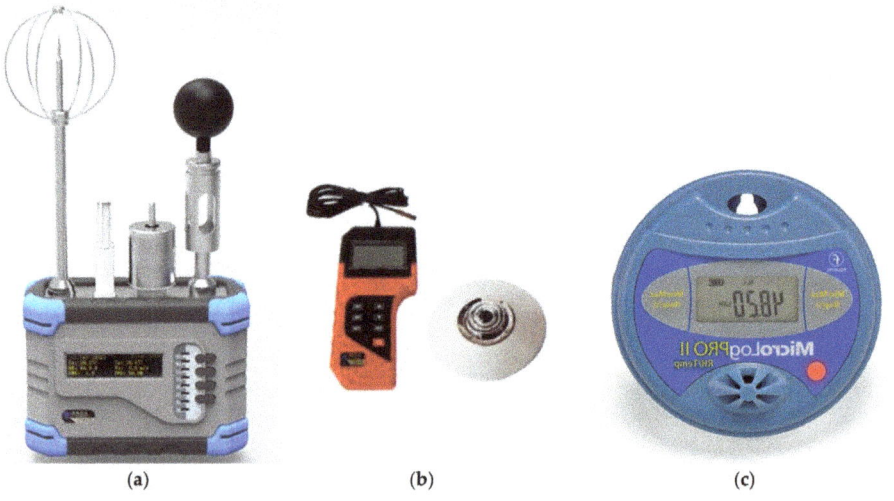

(a) (b) (c)

Figure 4. The measurement used during the study (**a**) JTSOFI-IAQ; (**b**) JTR05; and (**c**) Micro Log series.

For most of the subjects, the measuring instruments were positioned at 1.1 m above the floor for sitting positions, while for a few participants in standing positions, the measuring instruments were positioned at 1.6 m above the floor. In addition, outside air temperature was measured using the same measuring instruments.

3.3. Subjects

A total of 350 copies of a questionnaire were distributed in Danba County during winter, of which 327 were valid. The collected sample of survey participants included 138 male volunteers and 189 female volunteers. The age distribution of the participants ranged from 14 to 80 years of age. Most of the respondents were residents who had adapted to the living style and the local climate of the Danba County. The distribution of participant gender and age are presented in Table 4.

Table 4. Participant demographic information.

Sample	Type	Number (Percentage %)
Gender	Male	138 (42%)
	Female	189 (58%)
	Total	327
Age	14–26 years	(58) 17.74%
	27–38 years	(65) 19.88%
	39–51 years	(134) 40.98%
	52–80 years	(70) 21.40%
	Average	42 years

3.4. Thermal Comfort Model

3.4.1. PMV Index

PMV and PPD (predicted percentage of dissatisfied) assess thermal environments in air-conditioned buildings. The PMV index predicts comfort sensations on the ASHRAE's 7-point scale, while the PPD predicts the percentage of dissatisfaction about thermal environments. The PMV index is calculated with the following Equation [50]:

$$PMV = (0.303\, e^{-2100*M} + 0.028) * ((M - W) - H - E_c - C_{res} - E_{res}), \tag{1}$$

where M is the metabolic rate, in Watt per square meter (W/m^2); W is the effective mechanical power, in Watt per square meter (W/m^2); H is the sensitive heat losses; E_c—the heat exchange by evaporation on the skin; C_{res}—heat exchange by convection in breathing; E_{res}—the evaporative heat exchange in breathing. In Equation (1), the terms H, E_c, C_{res}, and H_{res}, correspond to the heat exchange between human body and surrounding environments.

As research indicates, there are two ways to calculate PMV values: tables and computer software tools [51]. In this study, particular, for the calculation of PMV index, ASHRAE Thermal Comfort Tool [52] was used, which determines PMV values based on ASHRAE-55 Standards. The input values for ASHRAE Thermal Comfort Tool include factors such as air temperature, mean radiant temperature, air speed, humidity, metabolic rate, and clothing level.

3.4.2. Operative Temperature

Binned indoor operative temperatures were determined for every 1°C, and were used to establish linear regression models and to determine the relationship between thermal comfort sensations and indoor temperatures. Regression coefficients and R square values were calculated and used to assess the power of linear regression equations. A confidence level of 95% was adopted in the statistical analysis in this paper to examine the strength of regression models. The calculation method of operative and the aPMV model is described in the following sections in more detail.

The indoor operating temperature (T_{op}) is calculated based on the average value of indoor air temperature (T_a) and indoor mean radiation temperature (T_r) assuming that occupant metabolic rates are between 1.0 met and 1.3 met, and air velocities are less than 0.20 m/s [53]. Investigation of the survey results showed that the indoor environmental conditions in the studied dwellings are suitable for calculating indoor operative temperatures with the above assumptions. Therefore, the indoor operative temperatures is determined using the following equation [54]:

$$T_{op} = (T_a + T_r)/2, \tag{2}$$

where T_{op} is indoor operative temperature, T_a is indoor mean air temperature, and T_r is indoor mean radiant temperature.

3.4.3. Adaptive Predicted Mean Votes (aPMV)

The PMV model of prediction of thermal sensations has been widely used in the past for the evaluation of thermal comfort conditions in buildings [55]. However, some discrepancies between PMV and actual mean votes (AMV) were observed in recent studies [56]. In fact, occupants tend to tolerate a broader temperature range than the range predicted by PMV models [57]. Accordingly, aPMV models based on black box methods have been developed by Yao et al. [8] to examine the relationship of PMV and AMV models in real environments considering occupant psychological and behavioural adaptations. The aPMV model is calculated as follows [8]:

$$aPMV = \frac{PMV}{1 + \lambda PMV}, \tag{3}$$

where λ is the adaptive coefficient that can be determined by actual field measurements affected by several factors, such as climate and culture. The Equation (3) can be called as the Adaptive Predicted Mean Vote model or aPMV model.

Research has shown that the body adapts to the local climate and environment [20]. The adaptation level can be measured by the adaptive coefficient λ obtained from Equation (3). The "λ" is the "adaptive coefficient" which was defined by Yao et al. [8]. In order to determine the λ, Equation (3) was rewritten as the following:

$$\lambda = y - x, \tag{4}$$

where $x = \frac{1}{PMV}$; and $y = \frac{1}{aPMV}$, in which x and y represent individual samples in the dataset (x_i and y_i). This indicates that the λ can be determined by least square methods using the following equation:

$$\prod = \sum_{i=1}^{n} [y_i - f(x_i)]^2 = \sum_{i=1}^{n} [y_i - (x_i + \lambda)]^2 = minimum \tag{5}$$

where $y_i = f(x_i)$; x_i and y_i represent ith data sample. In order to solve the above equation, the following equation could be used instead [54]:

$$\frac{\partial Q}{\partial a} = \sum_{2}(y_i - x_i - \lambda) = 0, \tag{6}$$

Therefore

$$\lambda = \frac{\sum_{i=1}^{n}[y_i - x_i]}{n}, \tag{7}$$

4. Results and Discussion

4.1. Annual Indoor Environmental Parameters

The relationship between the globe temperature (T_g) and air temperature (T_a) is presented in Figure 5. Air temperature is higher than the globe temperature for the studied period in winter. The average value of the difference between the globe temperature and air temperature is 0.95 °C, which can be explained by small window sizes and low lighting levels. Moreover, temperature ranges were less than 4 °C for both T_g and T_a. As mentioned in Section 3.4.1, the operative temperature is used instead of T_g and T_a in this study.

The outdoor air temperature ranged between −5.61 and 10.26 °C, and the indoor air temperature between 3.3 and 7.4 °C during the studied period in January. The relationship between the indoor and outdoor air temperatures is plotted in Figure 6. The trend-line of the outdoor and indoor temperature plot is shown in the Figure. The intercept of the equation is 5.93, which indicates that indoor air temperatures are almost 5.93 °C higher than the outdoor air temperatures. This is because the building envelope plays the principal role in preventing the heat escape from the building through building skin.

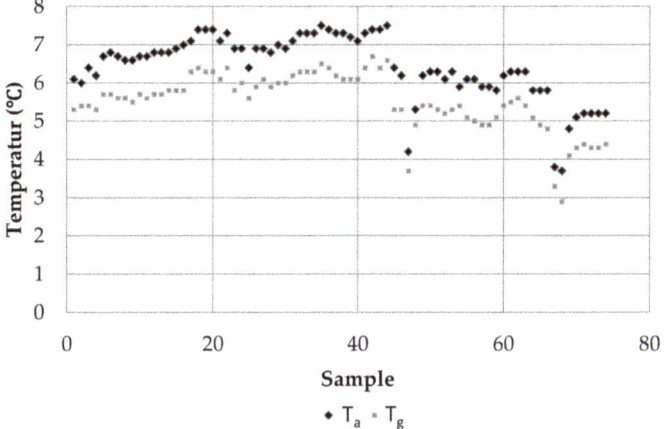

Figure 5. Comparison between globe temperature (T_g) and air temperatures (T_a).

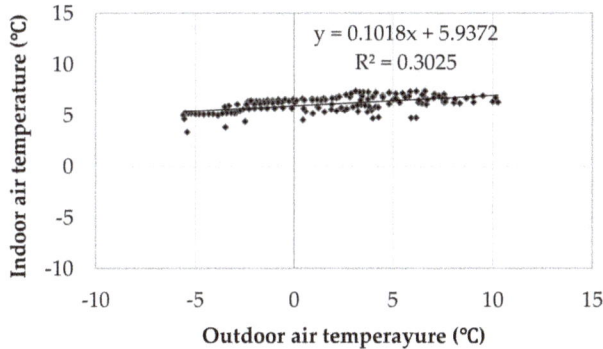

Figure 6. Relationship between indoor and outdoor temperatures in Danba.

The distribution of binned T_a is presented in Figure 7. From the figure, it can be observed that the frequency of the temperature between 6 and 7 °C is about 52%. Temperatures under 4 °C are only 1.3%. The distribution of the frequency of relative humidity (RH) is presented in Figure 8. As can be seen from the figure, the most frequent RH ranges were 47% to 55% with the value of 84% of the studied period.

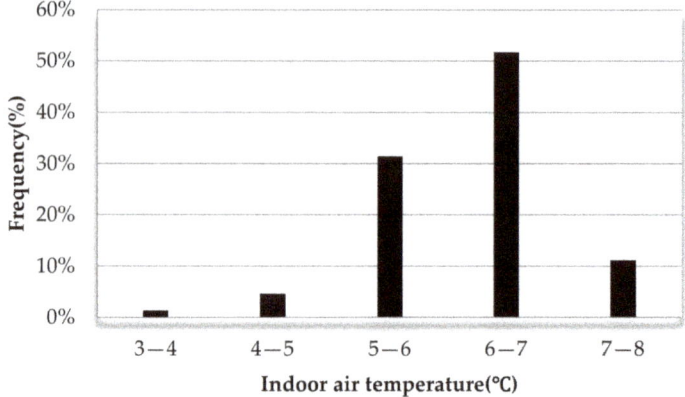

Figure 7. Frequency of the indoor air temperature in winter.

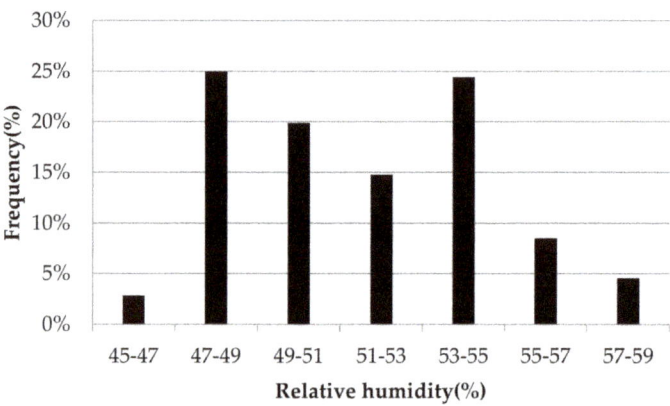

Figure 8. Frequency of the indoor air relative humidity in winter.

4.2. Adaptive Behaviours

The investigation of the adaptive behaviour question from the survey results revealed that Tibetan residents have a unique traditional lifestyle that adapts to their thermal environments. In the studied period, wearing more clothes, drinking hot high-calorie teas, and using a brazier were the most common adaptive behaviours that participants reported as their adaptive behaviours in winter (see Figure 9).

Particularly, drinking milk or buttered tea as high-calorie drinks, helps Tibetan residents not only to restore energy needs for the demanding Tibetan lifestyle but it also keeps them warm on cold winter days. The other popular adaptive behaviour was adjusting and adding clothing layers to minimise heat loss through an increased thermal resistance of clothing. Based on the survey results of adaptive behaviours, the most popular adaptive measure was adjusting clothing. The traditional Han clothing of warm jacket and cotton-padded shoes are popular outfits worn by Tibetan residents. Blocking house windows against cold air infiltration was the third most popular adaptive behaviour. Another observed adaptive measure in the stone dwellings was activity. Residents tended to increase their metabolism rate by increasing their activities. A linear correlation between clothing insulations and indoor temperature in winter is presented in Figure 10.

Figure 9. Adaptive behaviours in winter: (**a**) Dressing example in winter; (**b**) Small windows in stone houses; (**c**) Brazier; (**d**) Drinking milk or buttered tea.

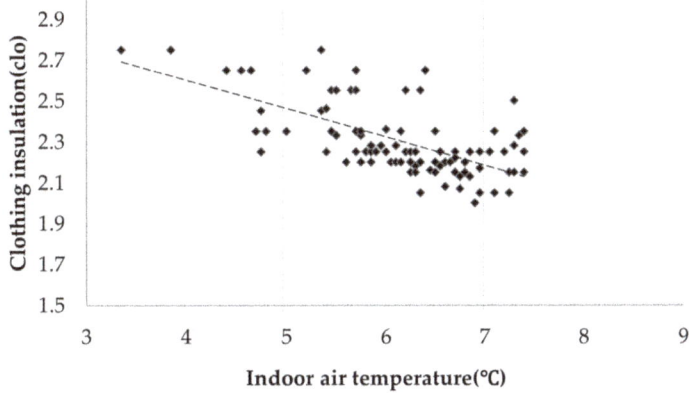

Figure 10. Correlation between resident clothing insulation and indoor air temperature in winter.

It was observed that clothing insulation levels were inversely proportional to the indoor air temperatures. When the indoor temperature was lower than 5 °C, there was a less significant linear relationship between clothing insulations and indoor air temperatures. When the indoor air temperature ranged between 5 and 7 °C, the fluctuation of clothing levels was more significant within the range between 2.0 clo to 2.2 clo.

4.3. Indoor Thermal Environment

The thermal sensation votes (TSV), humid sensation votes (HSV), and draft sensation votes (DSV) are shown in Figure 11. The percentage of votes for "neutral" and "slightly cooler" was 76%. The votes for "neutral" accounted for 29% (TSV = 0) and the votes for "slightly cool" accounted for 47% (TSV = −1). The proportion of people who experienced an acceptable humid sensation (HSV = −1, 0 or 1) was 91.01%, while the majority of participants (74%) felt slightly dry (HSV = −1). The percentage of people who experienced an acceptable draft sensation accounted for 93% (−1 to 1), and among those 64% felt neutral (DSV = 0). As a result, the indoor temperature scored only slightly cool and the air slightly dry in typical winter days.

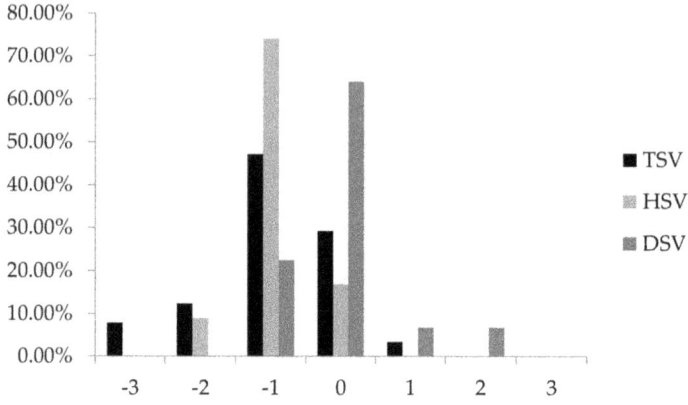

Figure 11. TSV, HSV, DSV in winter.

4.4. Comparative Analysis of PMV and AMV

By using regression analysis, the correlation of actual mean vote (AMV) and PMV with indoor operative temperature in winter was determined as follows:

$$AMV = 0.2018T_{op} - 2.606, (R^2 = 0.918) \tag{8}$$

where T_{op} is indoor operative temperature, which is determined based on Equation (2); AMV is the actual mean vote (see Figure 12). The neutral temperature (T_n) can be regarded as the temperature at which thermal sensation votes are "neutral". From Equation (8), the T_n from the AMV model in winter is 12.92 °C. The heat balance model of PMV equations for the studied period is as follows:

$$PMV = 0.1472T_{op} - 2.615, (R^2 = 0.984) \tag{9}$$

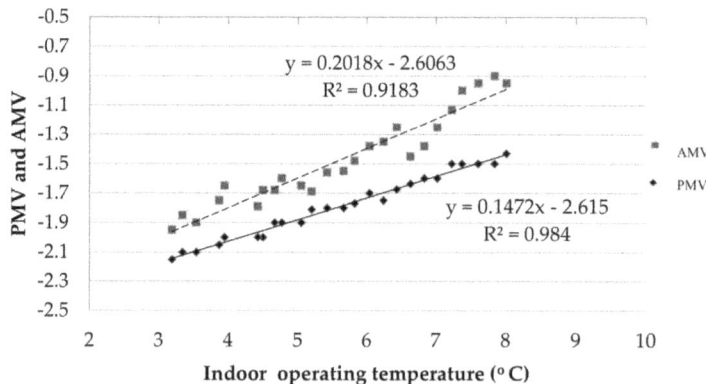

Figure 12. The linear relations between AMV, PMV, and indoor operating temperature (T_{op}) in winter.

The neutral temperature calculated based on the PMV model of Equation (9) is 17.76 °C, which is almost 5 °C higher than the neutral temperature of AMV model of actual sensation votes. This indicates that the actual neutral thermal sensations were much higher than the predicted thermal sensation by the PMV model. As Table 5 presents, the predicted thermal neutral temperature and the measured thermal neutral temperature have a deviation from the average operating temperature, and the average operating temperature is lower than that of the neutral thermal temperature in winter.

Table 5. Thermal neutral temperature statistics during winter.

Classification	Prediction	Temperature (°C)
Average operating temperature (°C)	-	6.29
Neutral temperature (°C)	Forecast	17.76
	Actual measurement	12.92

5. Adaptive Thermal Comfort

5.1. Obtaining the Adaptive Coefficient λ

In the studied samples, 25 groups were obtained with PMV value of less than 0 using the Bin method. Therefore λ can be determined by the following equation:

$$\lambda = \frac{\sum_1^{25}(y-x)}{25} = -0.32, \tag{10}$$

Thus, the aPMV model for a stone dwelling in Danba can be written as the following:

$$aPMV = \frac{PMV}{1 - 0.32 * PMV} \text{ (winter)}, \tag{11}$$

In order to compare the actual Predicted Mean Vote derived from the aPMV model and the actual thermal sensation vote (TSV), the aPMV and TSV were plotted in Figure 13. The aPMV model showed a stronger agreement with the PMV model than with the TSV model.

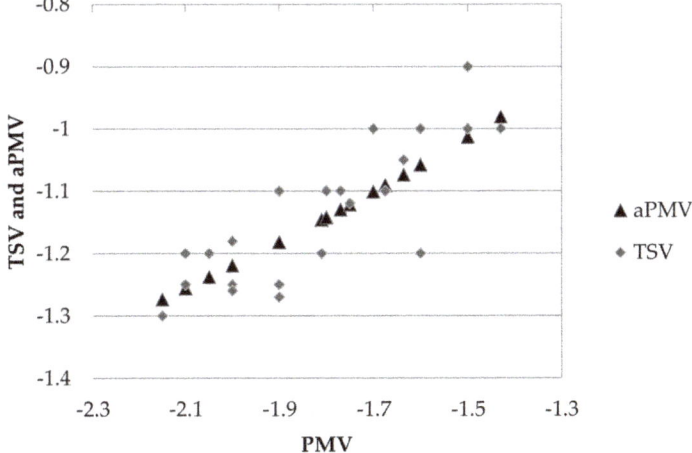

Figure 13. Correlation between aPMV and TSV.

5.2. Thermal Comfort Zone

Indoor air temperature, air humidity, air velocity, and radiation temperature are important factors that constitute indoor environmental conditions. These factors affect the health and body comfort of occupants. Based on the local weather data, the comfort zone is determined on the psychometric chart. The psychometric chart can also determine the heating or cooling potential of the passive design of a building based on local weather data.

The thermal comfort zone for residential buildings in Danba during winter is plotted in Figure 14 using the Climate Consultant Tool for the studied period. The abscissa represents the dry-bulb temperature (°C), the ordinate represents absolute humidity (g/kg), and the arc represents relative humidity (%). The enclosed area (yellow rectangle) in the figure represents the comfort zone, while the blue dots represent the data-points of environmental conditions during the survey period. As shown in the figure, none of the blue dots fall within the comfort zone, while in our survey, the majority of residents (76%) voted for an acceptable comfort sensation during typical winter days. This finding indicates that a revision of the comfort zone for Danba is needed with a reference to the local adaptive coefficient, which was calculated in this study. The Climate Consultant Tool assumes that the average radiant temperature is close to the air temperature and that the airflow velocity is within a comfortable range.

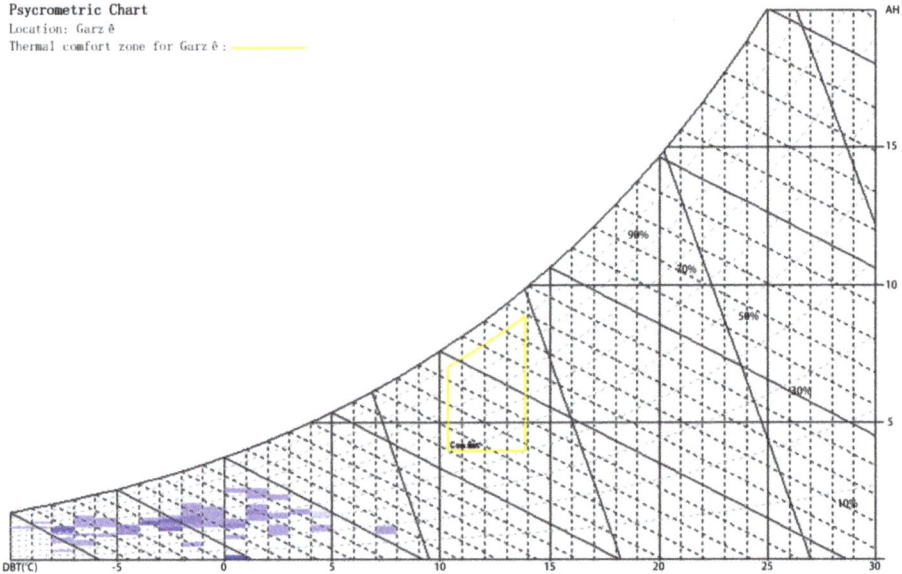

Figure 14. Thermal comfort zone for residential buildings in Danba during winter.

6. Conclusions

This study is based on subjective responses of 327 occupants in the cold climatic zones of Sichuan, China. The experiments were conducted in winter to evaluate the local residential levels of adaptations in cold conditions. The research findings are summarised as follows:

1. In winter, indoor air temperature in residential buildings fluctuates greatly from 3.1 to 8.1 °C, and the outdoor temperatures fluctuates between −5.61 and 10.26 °C, while the indoor relative humidity ranges from 54.3% to 70.8%. The indoor temperatures were higher than the outdoor temperature during the studied period.

2. The thermal neutral temperature (T_n) determined by the PMV model was almost 5 °C higher than the T_n observed based on the actual Mean Vote (AMV) model determined by the survey results. When AMV is equal to 0, the indoor thermal neutral temperature in winter was 12.92 °C, which was lower than the predicted thermal neutral temperature of 17.76 °C in this region.

3. The residents maintain a unique lifestyle in terms of thermal adaptation. They wear heavy clothing (2.0 to 2.2 clo) and drink milk or buttered tea to protect themselves from the cold winter. Another two popular adaptive behaviours are blocking windows against from draughts, and increasing activity.

4. It was observed that the PMV model underestimated the comfort sensations of the occupants in cold climates. Theoretical thermal comfort models have been developed that explain the deviation in the PMV and AMV plots.

5. An aPMV model and adaptive coefficient was developed for winter season in cold regions of China in Tibet.

6. By applying the adaptive thermal comfort model (aPMV model), the adaptive thermal comfort coefficient λ value of −0.32 has been obtained and validated for residential buildings in Danba.

This research also showed that the perception of thermal comfort depends on physiological and non-physiological factors influenced by culture, the availability of adaptive opportunities and culture. The developed adaptive coefficient could be used as a reference for thermal comfort evaluations in the cold regions with low-income residents.

Acknowledgments: This study has been funded by the Southwest University of Science and Technology PhD project "Research on green renewal of stone-built dwellings in Sichuan Tibetan Area" (grant number 16zx7133). Special thanks go to those who were involved and who helped in the field survey.

Author Contributions: All authors contributed equally to this paper.

Conflicts of Interest: The authors declare that there is no conflict of interest regarding the publication of this manuscript.

References

1. Mahdavi, A.; Kumar, S. Implications of indoor climate control for comfort, energy and environment. *Energy Build.* **1996**, *24*, 167–177. [CrossRef]

2. Gou, Z.; Lau, S.S.-Y.; Shen, J. Indoor environmental satisfaction in two LEED offices and its implications in green interior design. *Indoor Built Environ.* **2012**, *21*, 503–514. [CrossRef]

3. Gou, Z.; Lau, S.S.-Y.; Chen, F. Subjective and objective evaluation of the thermal environment in a three-star green office building in China. *Indoor Built Environ.* **2012**, *21*, 412–422. [CrossRef]

4. Khoshbakht, M.; Gou, Z.; Lu, Y.; Xie, X.; Zhang, J. Are green buildings more satisfactory? A review of global evidence. Early warning and assessment report series. *Habitat Int.* **2018**, in press. [CrossRef]

5. Khoshbakht, M.; Gou, Z.; Dupre, K. Energy use characteristics and benchmarking for higher education buildings. *Energy Build.* **2018**, *164*, 61–76. [CrossRef]

6. Khoshbakht, M.; Gou, Z.; Dupre, K. Cost-benefit prediction of green buildings: SWOT analysis of research methods and recent applications. *Procedia Eng.* **2017**, *180*, 167–178. [CrossRef]

7. Sun, X.; Gou, Z.; Lau, S.S.-Y. Cost-effectiveness of active and passive design strategies in existing building retrofits in tropical climate: Case study of a zero energy building. *J. Clean. Prod.* **2018**, *183*, 35–45. [CrossRef]

8. Yao, R.; Li, B.; Liu, J. A theoretical adaptive model of thermal comfort–Adaptive Predicted Mean Vote (aPMV). *Build. Environ.* **2009**, *44*, 2089–2096. [CrossRef]

9. De Dear, R.; Akimoto, T.; Arens, E.; Brager, G.; Candido, C.; Cheong, K.; Li, B.; Nishihara, N.; Sekhar, S.; Tanabe, S. Progress in thermal comfort research over the last twenty years. *Indoor Air* **2013**, *23*, 442–461. [CrossRef] [PubMed]

10. Fanger, P.O. *Thermal Comfort: Analysis and Applications in Environmental Engineering*; Danish Technical Press: Copenhagen, Denmark, 1970.

11. Alfano, F.R.A.; Olesen, B.W.; Palella, B.I. Povl Ole Fanger's impact ten years later. *Energy Build.* **2017**, *152*, 243–249. [CrossRef]

12. Xiong, Z.; Yang, K. Theoretical model of infrared radiation of dressed human body indoors. In Proceedings of the International Symposium on Photoelectronic Detection and Imaging 2007: Photoelectronic Imaging and Detection, Beijing, China, 3 March 2008; International Society for Optics and Photonics: Bellingham, WA, USA, 2008.

13. *ISO 8996. Ergonomics of the Thermal Environment e Determination of Metabolic Rate*; International Standardization Organization: Geneva, Switzerland, 2004.

14. *ISO 7726. Ergonomics of the Thermal Environment—Instruments for Measuring Physical Quantities*; International Standardization Organization: Geneva, Switzerland, 1998.

15. *ISO 10551. Ergonomics of the Thermal Environment e Assessment of the Influence of the Thermal Environment Using Subjective Judgement Scales*; International Standardization Organization: Geneva, Switzerland, 1995.

16. Fanger, P.O.; Toftum, J. Extension of the PMV model to non-air-conditioned buildings in warm climates. *Energy Build.* **2002**, *34*, 533–536. [CrossRef]

17. Alfano, F.R.A.; Ianniello, E.; Palella, B.I. PMV–PPD and acceptability in naturally ventilated schools. *Build. Environ.* **2013**, *67*, 129–137. [CrossRef]

18. Alfano, F.R.A.; Bellia, L.; Boerstra, A.; Van Dijken, F.; Ianniello, E.; Lopardo, G.; Minichiello, F.; Romagnoni, P.; da Silva, M.G. *REHVA-Indoor Environment and Energy Efficiency in Schools—Part 1*; REHVA: Brussels, Belgium, 2010; Volume 2010, pp. 1–121.

19. Dušan, K.; Miloslav, B.; Marián, V.; Anna, V.; Martin, K.; Maros, N. Analysis of an indoor environment in year-round operation. *Adv. Mater. Res.* **2013**, *649*, 37–40.

20. De Dear, R.J.; Brager, G.S.; Reardon, J.; Nicol, F. Developing an adaptive model of thermal comfort and preference/discussion. *ASHRAE Trans.* **1998**, *104*, 145.

21. Humphreys, M.A.; Nicol, J.F.; Raja, I.A. Field studies of indoor thermal comfort and the progress of the adaptive approach. *Adv. Build. Energy Res.* **2007**, *1*, 55–88. [CrossRef]

22. Brager, G.S.; De Dear, R.J. Thermal adaptation in the built environment: A literature review. *Energy Build.* **1998**, *27*, 83–96. [CrossRef]

23. De Dear, R.J.; Brager, G.S. Thermal comfort in naturally ventilated buildings: Revisions to ASHRAE Standard 55. *Energy Build.* **2002**, *34*, 549–561. [CrossRef]

24. Arif, M.; Katafygiotou, M.; Mazroei, A.; Kaushik, A.; Elsarrag, E. Impact of indoor environmental quality on occupant well-being and comfort: A review of the literature. *Int. J. Sustain. Built Environ.* **2016**, *5*, 1–11.

25. Yao, R. Indoor Climate Simulation and Thermal Comfort Research. Ph.D. Thesis, Chongqing University, Chongqing, China, 1997.

26. Cao, B.; Zhu, Y.; Ouyang, Q.; Zhou, X.; Huang, L. Field study of human thermal comfort and thermal adaptability during the summer and winter in Beijing. *Energy Build.* **2011**, *43*, 1051–1056. [CrossRef]

27. Han, J.; Zhang, G.; Zhang, Q.; Zhang, J.; Liu, J.; Tian, L.; Zheng, C.; Hao, J.; Lin, J.; Liu, Y. Field study on occupants' thermal comfort and residential thermal environment in a hot-humid climate of China. *Build. Environ.* **2007**, *42*, 4043–4050. [CrossRef]

28. Yang, L.; Yan, H.; Xu, Y.; Lam, J.C. Residential thermal environment in cold climates at high altitudes and building energy use implications. *Energy Build.* **2013**, *62*, 139–145. [CrossRef]

29. Zhang, Y.; Chen, H.; Meng, Q. Thermal comfort in buildings with split air-conditioners in hot-humid area of China. *Build. Environ.* **2013**, *64*, 213–224. [CrossRef]

30. Zhang, Y.; Wang, J.; Chen, H.; Zhang, J.; Meng, Q. Thermal comfort in naturally ventilated buildings in hot-humid area of China. *Build. Environ.* **2010**, *45*, 2562–2570. [CrossRef]

31. Wang, Z.J.; Wang, G.; Lian, L.-M. A field study of the thermal environment in residential buildings in Harbin. *Trans. Am. Soc. Heat. Refrigerating Air Cond. Eng.* **2003**, *109*, 350–355.

32. Zou, M.; Yang, L.; Yan, H. Investigation of summer thermal comfort in village and town house in Hanzhoung. *Build. Sci. Res. Sichuan* **2013**, *39*, 341–345.

33. Yan, H.; Yan, L.; Zhou, B. Field research on indoor thermal comfort of residential buildings in Jiaozuo in winter. *J. HV&AC* **2011**, *41*, 119–125.

34. Yang, L.; Yang, Q.; Yan, H.; Liu, J. Investigation and Study on indoor thermal comfort of rural residential buildings in Guanzhong, Shaanxi. *J. Xi'an Univ. Archit. Technol.* **2011**, *43*, 551–556.

35. Li, B.; Zheng, J.; Yao, R.; Jing, S. *Indoor Thermal Environment and Human Thermal Comfort*; Chongqing University Press: Chongqing, China, 2012.

36. Zhou, L.; He, Y.; Li, N.; Zhang, W. Investigation on thermal comfort of rural residential buildings in winter in Xiangxi. *Build. Sci.* **2016**, *32*, 29–33.

37. Wang, J.; Wang, H.; Liu, Y. The thermal comfort of the students in the primary and secondary schools in Qinghai. *J. Civ. Archit. Environ. Eng.* **2017**, *39*, 32–37.

38. Li, K. Evaluation of Indoor Thermal Environment in Rural Housing in Chongqing. Ph.D. Thesis, Chongqing Univercity, Chongqing, China, 2015.

39. Yang, Z.; Xu, Y.; Peng, M. Adaptability of wood wall dwellings to hot and humid climate. *J. Civ. Archit. Environ. Eng.* **2016**, *38*, 1–6.

40. Zheng, W.; Yang, L.; Song, X. Climatic seasonal changes of adaptive level of human body in cold regions. *J. HVAC* **2015**, *2015*, 80–85.

41. Wang, Y. Study on the Characteristics of Indoor Thermal and Wet Environment in Tibet. Ph.D. Thesis, Chongqing University, Chongqing, China, 2015.

42. He, Q.; Liu, D.; Zhu, X. Test and study on the indoor thermal environment of Tibetan folk dwellings in Western Sichuan Plateau. *J. Xi'an Archit. Technol.* **2015**, *3*, 402–406.

43. Ou, J.; Wang, C.; Li, Q. Investigation and measurement of the heat environment of the winter ledger in the pastoral area of Western Sichuan Plateau. *Build. Sci.* **2016**, *6*, 39–44.

44. Chen, Y.; Chen, B.; Gao, M. Indoor thermal and wet environment test of a Tibetan stone dwelling in Danba during the transition season. *Sichuan Archit.* **2017**, *37*, 63–64.

45. Gou, Z.; Xie, X.; Lu, Y.; Khoshbakht, M. Quality of Life (QoL) Survey in Hong Kong: Understanding the Importance of Housing Environment and Needs of Residents from Different Housing Sectors. *Int. J. Environ. Res. Public Health* **2018**, *15*, 219. [CrossRef] [PubMed]

46. McCullough, E.A.; Jones, B.W.; Huck, J. A comprehensive data base for estimating clothing insulation. *ASHRAE Trans.* **1985**, *91*, 29–47.

47. *ASHRAE Standard 2004. Thermal Environmental Conditions for Human Occupancy;* American Society of Heating, Refrigerating and Air-Conditioning Engineers: New York, NY, USA, 2004.

48. Alfano, F.R.A.; Palella, B.I.; Riccio, G. The role of measurement accuracy on the thermal environment assessment by means of PMV index. *Build. Environ.* **2011**, *46*, 1361–1369. [CrossRef]

49. Cao, B.; Luo, M.; Li, M.; Zhu, Y. Too cold or too warm? A winter thermal comfort study in different climate zones in China. *Energy Build.* **2016**, *133*, 469–477. [CrossRef]

50. da Silva, M.C.G.; Pires, J.N.; Loureiro, A.; Pereira, L.D.; Neto, P.; Gaspar, A.; Viegas, D.X.; Soares, N.; Oliveira, M.; Costa, J. Spreadsheets for the calculation of Thermal Comfort Indices PMV and PPD. *Renew. Sustain. Energy Rev.* **2014**, *40*, 911–922.

51. Alfano, F.R.A.; Palella, B.I.; Riccio, G. Notes on the Calculation of the PMV Index by Means of Apps. *Energy Procedia* **2016**, *101*, 249–256. [CrossRef]

52. Fountain, M.; Huizenga, C. *A Thermal Sensation Prediction Software Tool for Use by the Profession;* American Society of Heating, Refrigerating and Air-Conditioning Engineers: New York, NY, USA, 1997.

53. International Organization for Standardization. *Ergonomics of the Thermal Environment: Analytical Determination and Interpretation of Thermal Comfort Using Calculation of the PMV and PPD Indices and Local Thermal Comfort Criteria;* International Organization for Standardization: Geneva, Switzerland, 2005.

54. Yu, W.; Li, B.; Yao, R.; Wang, D.; Li, K. A study of thermal comfort in residential buildings on the Tibetan Plateau, China. *Build. Environ.* **2017**, *119*, 71–86. [CrossRef]

55. Kim, J.T.; Lim, J.H.; Cho, S.H.; Yun, G.Y. Development of the adaptive PMV model for improving prediction performances. *Energy Build.* **2015**, *98*, 100–105. [CrossRef]

56. Nicol, J.F.; Humphreys, M.A. Adaptive thermal comfort and sustainable thermal standards for buildings. *Energy Build.* **2002**, *34*, 563–572. [CrossRef]

57. Xu, W.; Chen, X.; Zhao, J. An adaptive Predicted Mean Vote (aPMV) model in office. In Proceedings of the Mechanic Automation and Control Engineering (MACE), Wuhan, China, 26–28 June 2010.

© 2018 by the authors. Licensee MDPI, Basel, Switzerland. This article is an open access article distributed under the terms and conditions of the Creative Commons Attribution (CC BY) license (http://creativecommons.org/licenses/by/4.0/).

Article

The Design of Local-Authority Rental Housing for the Elderly That Improves Their Quality of Life

Yukiko Kuboshima *, Jacqueline McIntosh and Geoff Thomas

School of Architecture, Victoria University of Wellington, Wellington 6011, New Zealand;
Jacqueline.McIntosh@vuw.ac.nz (J.M.); Geoff.Thomas@vuw.ac.nz (G.T.)
* Correspondence: Yukiko.Kuboshima@vuw.ac.nz

Received: 30 March 2018; Accepted: 11 May 2018; Published: 16 May 2018

Abstract: As the population ages, the demand for suitable rental housing will increase. Suitable housing means housing that can accommodate those impairments that typically correspond with ageing. This paper explores the quality of life (QoL) requirements of those elderly with high-care needs who live in rental housing. Using a qualitative case study approach, it examines the living experiences of six elderly people who need assistance and are living in local-authority rental housing in New Zealand. The themes of QoL were identified from the literature and related to the larger themes of; 1. Activities and independence, 2. Sense of control, 3. Privacy, 4. Relationships, 5. Quality of care, and 6. Comfort. The survey consisted of a detailed documentation of the physical environment, followed by interviews with and full-day observations of the residents and their caregivers. The study finds that the design of housing that improves their QoL requires solutions to accommodate the various conflicting needs for their QoL that include those derived from the diversity in the user's preferences and impairments. In the design of rental housing, there is greater need for additional or reorganized space to accommodate caregivers and visitors, maintain residents' independence, privacy, and other aspects important for their QoL.

Keywords: elderly; quality of life; housing design; rental housing; high care needs; post occupancy evaluation; qualitative research

---

1. Introduction

The ageing population is increasing both in the proportion and in the number of people over 65 living in New Zealand, as well as many other developed countries. As people age, they have a greater propensity for impairment and difficulty performing everyday tasks. In addition, psychological concerns such as insecurity and loneliness also become more prevalent [1,2]. At some point, typically in their 70s or later, these factors may induce them to seek a more suitable dwelling [3]. Some consider moving closer to their children; however, most New Zealanders prefer to avoid 'being a burden' [4]. To accommodate the projected rapid growth in the older people with high care needs [5], there is an increasing need for housing that supports those who require assistance to live independently. 'Ageing in place' can provide a greater quality of life (QoL) for the elderly [4] and for this reason it is promoted by the New Zealand government and internationally [6], increasing the demand for housing that enables those elderly with high-care needs to live as independently as possible.

Currently, in New Zealand, there are three main types of housing that provide some level of care and assistance to the elderly; retirement villages, public-sector housing (central government housing and local-authority housing), and private-sector rental housing, which includes community providers and religious and charitable groups. While retirement villages offer company and security and reduce concerns about home maintenance and care, they are only viable options for homeowners and those relatively well-off [7]. The demand for rental housing by the ageing population is projected to increase rapidly, influenced by the decline of levels of homeownership [8–11]. In response, recent government initiatives are seeking to address this situation, by encouraging community housing sectors to grow [12].

Rental housing tends to be less suitable for those elderly with disabilities, in terms of the provision of care and support, and access and facilities for the disabled when compared with housing in retirement villages [13]. While a high proportion of residents receive personal care in rental housing, the current rental housing generally fails to accommodate those with higher levels of dependency [14]. There is a growing demand to provide physical environments that facilitate the high-needs elderly to live independently in rental housing.

The quality of life of high-needs older people has been studied by many researchers [15,16]. Common themes include needs for independence, activities, relationships, identity, and quality of care. With increased needs for assistance, privacy is also an issue for those receiving care, both in a facility and at home [1,17]. In addition to the general effects of ageing, it has been reported that living in rental housing has a negative impact on resident QoL [18–20]. For example, they have less autonomy in making modifications to their dwelling to make it suitable for their use, compared with homeowners [18]. With these two factors in mind, there should be careful consideration of the design of rental housing in order to provide the high-needs elderly with greater QoL.

Post Occupancy Evaluation (POE) refers to the assessment of how buildings are used to support buildings' environmental performance and occupant wellbeing and productivity [21,22], which 'provides evidence of a wide range of environmental, social and economic benefits core to sustainability [22] (p. 6)'. POE has been conducted in various building types, commonly on public buildings such as schools, but on housing as well [23,24]. Increasingly the significance of POE has been recognized internationally. For example, POE has been introduced into the curriculum of many architecture schools in the UK [25] and the Institute of British Architects called on all central government-funded projects to commit to POE in 2017 [26]. In New Zealand, POE has been conducted for public educational buildings since 2015 [27]. It has also been reported that design professionals' interest in POE is growing in large architecture firms in the US and Canada [28]. While systematic approaches have been sought, a range of approaches and methods are currently used to carry out POE, depending on the focus of the particular outcome and range from data monitoring and quantitative evaluation to the qualitative surveys of occupants [22].

This paper focuses on QoL for the high-needs elderly who live in local-authority rental housing specially built for the elderly in the Wellington region in New Zealand. This research examines how the physical environment supports or undermines their independent life with high QoL, which provides evidence and directions for the effective design of rental housing for a sustainable society with a growing ageing population.

2. Method

An ethnographic case study was conducted for six elderly people who need assistance in daily life and are living in senior housing complexes in the Wellington Region of New Zealand. Ethnography is a qualitative method that has a great deal of potential in post-occupancy studies and many uses in architecture and the built environment [29]. Inclusion criteria were those who received assistance from a professional caregiver, were more than 70 years of age, and were interested in participating in the survey. Participants were selected through a questionnaire for the elderly which formed part of a prior study of housing options for those with high care needs. The survey consisted of: 1. documentation of the physical environment of the house; 2. semi-structured interviews with the elderly residents and their caregivers; and 3. personal observations of the residents during a full day, including unstructured interviews and informal conversation. Ethics approval was obtained from Victoria University Human Ethics Committee [Approval number: 23243]. For the focus of this study, six elderly people who lived in local-authority rental housing were selected for analysis (Table 1). Through the analysis of transcribed interviews and observation notes, the themes that emerged were coded in relation to broader QoL themes. For each theme, similarities and differences among the cases and the reasons of them were analyzed and relevant design themes were identified. Integrating the results and the design considerations of housing for the elderly that improve their QoL are discussed.

3. Themes for QoL

Through analysis, sub-themes were identified and related to the larger themes of; 1. Activities and independence, 2. Sense of control, 3. Privacy, 4. Relationships, 5. Quality of care and 6. Comfort, which is described in this section.

3.1. Activities and Independence

3.1.1. Circulation and Space for Movement

All residents wanted barrier-free environments. They experienced difficulties in moving with their walker frames (Residents 1, 5) and felt pain when going up and down the steps (Resident 3). Residents 1, 5, and 6, who did not always use an aid indoors, needed to hold walls, furniture and fixtures to maintain balance while walking. Resident 2, who was dependent on a trolley for support with walking, had to be very careful when moving with hot drinks or soup on the trolley over the connection between different floor materials, some of which had less than a centimeter level difference. Putting low tables in the middle of the room could result in a fall (Resident 6). There should be enough space both for walking and for furniture layout along the wall.

For outdoor mobility, Resident 1 had trouble going down three steps with a walker frame to the roadway to get on a taxi. Resident 2 experienced difficulty passing through the exterior swing door with a threshold while holding the trolley, while Resident 6, who could hold the walker frame over the same level difference, did not have trouble in getting in and out. Resident 5 had a sliding door at the unit entrance; however, there was a big level difference just out of the door, which was difficult for the resident to step out with a walker frame. To mitigate the inconvenience, a step (a brick) was installed to fill the level difference.

Vehicles were also used by some residents. Resident 2 used a mobility scooter, which was stored in the unit. To reach the scooter, the resident had to use the trolley to get close enough to transfer from the trolley to the scooter, then turn the scooter around to exit. Resident 3, used a bicycle, but had to store it on an exposed concrete deck. Accessible under cover storage is desirable for these mobility aids.

There were differences in the suitability of the physical environment depending on the types/levels of impairment, which, in turn, related to requirements for aids/vehicles for moving indoors and outdoors. The differences included the need to hold walls when walking and the extent to which barrier-free interventions were needed. Corridor and passageway widths should be considered in terms of walls and furniture required to support walking. Design for accessibility should be carefully considered in the door design, as well as floor design in both interior and outdoor spaces.

Table 1. Basic information on residents and settings.

Cases	Case 1	Case 2	Case 3	Case 4	Case 5	Case 6
Resident						
Age group	80–84	70–74	75–79	75–79	85–89	85–89
Levels of dependency	High (DS [1] = 7)	High (DS = 4)	Low (DS = 2)	Low (DS = 1)	High (DS = 4)	Low (DS = 0)
Types of impairment	Arthritis (legs) Had operations in knees and shoulders.	Post stroke (has left hemiplegia, the right hand shakes) Pain/arthritis (shoulder, back)	Diabetes Pain/arthritis (knees) Had hip replacement Pancreatitis	Spinal conditions Pulmonary conditions Pain/arthritis (shoulder)	Pain/arthritis (legs/knees)	Arthritis (legs) Cataracts Post stroke (currently few aftereffects remain)
Required assistance	Bathing Dressing Personal hygiene Put on/off compression stockings Household tasks	Bathing Personal hygiene Put on/off compression stockings Household tasks	Bathing Personal hygiene Household tasks	Bathing Household tasks	Bathing Dressing Household tasks	Household tasks
Mobility aid	Walker frame (in/outdoors)	Trolley (indoors), mobility scooter (outdoors)	Walker frame and bicycle (outdoors)	None	Stick (indoor), walker frame (in/outdoors)	Walker frame (in/outdoors)
Years of residence	26	9	9	7	17	7
Setting						
Unit type [2]	Bedsit B	Bedsit A	One-bedroom	Bedsit A	One-bedroom	One-bedroom
Building type	Apartment (outdoor access)	Semi-detached (outdoor access)	Semi-detached (outdoor access)	Apartment (outdoor access)	Apartment (outdoor access)	Semi-detached (outdoor access)
Unit size	32 m²	35 m²	41 m²	32 m²	45 m²	42 m²
Unit plan [2]						
Communal room	No	Yes	No	Yes	No	Yes

[1] DS ('Dependency Score'): For each person, six activities (bathing, dressing, personal hygiene, moving indoors, moving from bed to wheelchair/chair and eating) were given the scores of 0–2 or 0–3, according to the degree of assistance they receive. They were given a 'Dependency Score,' the total of the six scores for each activity. [2] Bedsit units are divided into two types; the bed area and the lounge being separated by the curtain (A) and otherwise (B). [2] Abbreviations used in the plans stand for; L: lounge, BR: bedroom, BS: bedsit, B: bathroom, K: kitchen.

3.1.2. Spaces for Sitting and Various Activities

All participants sat most of the time on a chair in their lounge/bedsit space, except for one resident, who spent his time mostly in his bedroom lying on a bed or sitting at desks (Resident 5). Four of them used an armchair, which allowed them to rest in their most comfortable posture and to adjust their leg and back position. Resident 4 mainly sat on an unpadded side chair at the table in the lounge part of his bedsit, which allowed him to sit straight, and by doing so he could avoid spinal pain.

While there were common activities for all in the sitting space such as watching TV, other activities varied depending on preferences. For example, Residents 2, 3, 4, and 5 liked playing games or messaging with their PC, while Residents 1 and 6 liked knitting, reading and/or crosswords, for which the space was used differently. There were also differences in activities, depending on the level/type of impairment. Residents 2, 4, and 5 sat at the PC desk for playing computer games or emailing/online chatting. Sitting at the desk was particularly necessary for Resident 2 to support a paralyzed left arm, while Resident 3 could operate his laptop putting it on the armrest of the armchair. Some residents also liked to see outside, which gave them chance to talk to neighbors (Residents 2, 4, and 6). Feelings of keeping occupied were important for them. Resident 3 could carry out multiple activities while sitting in the armchair such as looking at a PC that was put on the armrest for playing games while hearing was mostly attuned to the TV. For the various activities to be carried out from the sitting space, there were level surfaces within their reach such as tables, a kitchen bench, or shelves (including those under a kitchen bench and those of a trolley). These were essential to accommodate various things such as glasses, phones, remote controls, medicine, cups of tea, mail, pens, and papers.

The spatial organization of sitting spaces should allow a layout with an armchair and immediately adjacent tables and shelves to keep things within reach. The design of these spaces should also facilitate residents' various activities including watching TV. In particular, space that accommodates a table as well as a chair is necessary for high-dependency elderly with limited posture options.

3.2. Sense of Control

3.2.1. Ease of Maintenance, Keeping Space Clean and Tidy

It was important for residents to keep their spaces clean and tidy to maintain their sense of control. Resident 3 had many shelves at various heights within reach, which were very useful. However, Residents 1, 2, 4, and 6 did not have enough shelves and filled an adjacent table with necessary things; two of them did so in a less organized way as well as on the floor near their chair. They wanted more shelves at an appropriate height near their chair. Higher shelves were rarely used by those with higher levels of impairment, because they could only reach the front of the shelf and could not use a step ladder to reach the rear area of the shelf. Resident 2, whose hands shook, often spilt liquids (tea/soup) and did not like carpet, which stained easily and was never cleaned, even by the home-helper.

Difficulties in keeping the space clean and tidy varied depending on the types/levels of impairment. There should be consideration with regard to the interior elevation that provides shelves and storage of appropriate height and depth. Greater consideration of maintenance and cleaning with respect to floor materials is also required.

3.2.2. Control over Visitors

All six participants experienced a sense of control when they could see visitors were coming before they actually arrived. Each had a view of the doorway from their sitting space; however, there were differences in the extent to which the view of the visitors was restricted before they actually arrived, depending on the spatial layout. Three residents had lounge spaces facing the front of the dwelling with a view to a long driveway and liked that they could see who was coming (Residents 2, 3, and 6). Particularly Residents 3 and 6, who had enough distance between the unit and the pathway could have time to mentally prepare for having guests. However, Resident 1, whose lounge did not face the front, could only see who was coming through the window next to the front door at a distance just before they arrived. Residents 1 and 2, who had limited mobility, often invited visitors in while remaining seated, calling out a greeting and invitation to 'Come in.' Resident 5, who spent their time mostly in the bedroom, could not see nor hear the visitor arrive. For those with mobility issues, there should be a clear line of sight from the sitting space to the door, as well as sufficient proximity for a visitor to hear their greeting and welcome through the door.

3.3. Privacy

3.3.1. Privacy against Passers-By

When there was insufficient space between the dwelling unit and pathway, some residents felt a loss of privacy (Resident 1) and would shut the curtain (Resident 2) because 'people can easily look inside'. This was not a worry of others such as Residents 3, 5, and 6 who had no path nearby where many people could pass by. Privacy concerns are particularly important in small dwellings where many people can pass by in close proximity. Resident 4 also criticized the path layout in his site that allowed public access through the site late at night, despite the notice of 'No Public Access'.

3.3.2. High Privacy Needs for Incontinence

Resident 2 had high privacy needs related to incontinence. When the lead researcher was situated near the sitting space for the observation part of the survey, the resident tried to hide and pass water in the next room (approx. 3 m away); however, they were unable to reach the privacy of this room in time and passed water near the armchair. Resident 3 also mentioned frequent toileting at night (every 2 h); however, this did not impact on privacy. There should be consideration in the spatial design of spaces for highly dependent people that meet the special needs for privacy related to issues such as incontinence.

3.4. Relationships

3.4.1. Socializing through Communal Activities

Residents 1, 4, and 6 were fond of socializing and maintaining relationships with others; they actively engaged in various kinds of social activities such as personal hobby groups and social gatherings in council housing complexes. One of them wished for a community room in their own site. In contrast, Resident 2 did not attend any communal activities organized for residents because of concerns with incontinence. Residents 3 and 5, who had no organized activities nor any communal space in the complex where they lived, did not wish for them, because they preferred to keep in touch with other residents more personally. Differences in the manner of socializing with others can be affected by impairment as well as personal preferences.

3.4.2. Space for Having Guests

All residents had visits from families and friends, which was important for them. Residents 1 and 4, who had only a bedsit space, did not have opportunities to have guests stay the night. One of them wished for a separate lounge because the room was 'more like a bedroom'. Resident 2 wished for a solid wall rather than a curtain between the sitting space and the bed space for improved privacy. Resident 3 living in a one-bedroom unit liked the layout of their space with a dedicated private bedroom. Resident 5 used the lounge only when guests visited, but he appreciated having it. Resident 6 liked the idea of having a second bedroom for a visitor to stay. To accommodate a larger number of day-time visitors, Resident 6 used the open lawn space leading out from her lounge.

3.5. Quality of Care

3.5.1. Spaces for Assisted Showering

Five residents had a bathroom with a step at the entrance. (Resident 6 had a shower area with no level difference, and did not require assistance in showering.) Residents 1 and 4 had a bathtub with an overhead shower attached to the wall. These factors increased their caregiver's labor when assisting the mobility-impaired residents to bend forward and draw water with a bucket, as well as increasing the resident's risk of falling. Residents 3, 4, and 5 had a small, enclosed shower booth with a step at the entrance (less than 1-m square). This was not preferred by the caregiver because it was not big enough for her to go in to assist with washing. Bathroom size was also problematic for Resident 1 as it did not have enough space for drying with a caregiver's assistance. For assisted showering, sufficient space is required for a caregiver both for washing and for drying off.

3.5.2. Independence and Privacy in Assisted Showering

The amount of assistance required for showering varied by level of impairment. All residents wanted to do as much as possible themselves during showering to keep their independence and privacy. Resident 3, who only needed assistance in washing their legs and back and in drying, undressed themselves by their armchair before the caregiver arrived, then washed themselves in the shower with the curtain closed before requesting assistance, to maintain privacy. Resident 5 also shut the curtain during washing except for when he required assistance in washing his back. Their caregiver said that it was important 'for their dignity.' Resident 1, who required assistance in every activity associated with showering except for undressing, could have had greater independence and privacy if the shower type was not the one attached to the wall. A detached hose-type shower could have allowed them to wash private areas independently.

Special consideration of the fittings, furnishings and fixtures in the shower area is required for elderly people with mobility impairments. In addition, consideration should be given to the design of showering areas so as to allow the caregiver to keep out of the sight of residents for their privacy and dignity. The proximity of the space used for undressing to the bathroom is also important for improving privacy.

3.6. Comfort

Warmth and the Sun

Residents 1, 2, and 5 felt their units were cold. Particularly Resident 5 complained about coldness, even when using his own electric heater, which encouraged him to stay in the bed rather than sit. On the contrary, Resident 3, who had a heat pump, did not feel cold. Resident 1's caregiver commented on the importance of insulation in the walls. A carpet was preferred by many residents to a vinyl floor, because it was warmer (Residents 1, 3, 4, 6).

Sunshine was important for warmth as well as brightness for all residents; however, access to sunlight varied due to both issues related to their impairments as well as spatial design. Resident 2 opened the curtains only when it was sunny, because of high privacy needs resulting from health concerns. Resident 6, who had an issue with eyesight, had to be careful not to expose their eyes to the sunlight. Glare and reflection on the TV and PC screens limited access to natural light for Resident 3, who found it necessary to shut one of the curtains during the daytime. In the design and placement of windows, there is a need to meet both enhanced requirements for privacy as well as controlled access to sunlight to limit glare on TV/computer screens. The trees near resident's units and unit layout also affected sunlight. Resident 4 complained that tall trees planted in the north aspect from the unit blocked the sun and desired smaller trees. Resident 5, whose unit faced the west, also desired to get more sun, which was blocked by other blocks of three-story apartments. On the contrary, Resident 1 could fully enjoy the sun without any obstacles. The porch facing the north was used by one resident to sit in the sun in summer (Resident 1).

4. Design Considerations

For each sub-theme for QoL, similarities and differences for three cases were analyzed qualitatively and the relevant design issues for each of the themes for QoL were distilled and summarized in Table 2.

Table 2. Analysis of themes for Quality of Life (QoL) and relevant design themes.

Themes for QoL	Similarities	Differences [1]	Design Themes
Independence and activities			
Circulation and space for movement	Desire for environments with no level changes	(I) Assistance requirements for walking (I) Suitable storage space for mobility vehicles (P) Walkability	Spatial organization (Interior/exterior) Floor design Door design
Spaces for spending most of the time and various activities	Preference for spending most of the time sitting or lying	(I) Type of chair (P) Kinds of space required for varied activities (I) Required seating configuration for activities	Spatial organization (interior)
Sense of control			
Ease of maintenance, keeping space clean and tidy	Preference for shelves of appropriate height and depth	(I) Preferences for floor finishes (floors to be cleaned easily). (I) Ability to access to storage	Flooring design Storage design Interior elevation design
Control over visitors	Preference for a view to the door from their sitting space	(I) Requirements for the view and proximity to the door	Spatial organization (interior/exterior) Exterior elevation design
Privacy			
Space for privacy against passers-by	-	-	Spatial organization (interior/exterior) Exterior elevation design Path layout
High privacy needs for incontinence	-	(I) Degrees of privacy needs depending on the health issue	Spatial organization (interior/exterior)
Relationships			
Socializing through communal activities	-	(I,P) Ways of socializing with others	Communal space
Space for welcoming visitors	Preference for the separation of bedroom from the lounge	(P) Number of visitors and manners of accommodating guests	Spatial organization (interior/exterior)
Quality of care			
Space for assisted showering	Need for an accessible shower area and space for caregivers to assist washing and drying	-	Size of space (shower area) Types of shower and the area
Independence and privacy in assisted showering	Wish to do as much as they could by themselves	(I) The amount of the assistance required	Equipment/fixture Spatial organization (shower area)
Comfort			
Warmth and the sun	Preference for warmth and the sun	(I,P) The degree of sunlight preferred	Spatial organization (interior/exterior) Exterior elevation design Floor design Insulation, heater Unit layout

[1] (I): differences by types/levels of impairments, (P): by preferences.

Considerate design of interior space in individual units and adjacent facilities with regards to exterior space can improve the QoL for older people with restricted mobility. Important design considerations are discussed for each design theme, integrating the information obtained through the analysis.

4.1. Consideration for Accommodating Various Levels/Types of Impairments and Preferences

The design requirements for greater QoL are affected by the type and level of impairments. There were differences in the design requirements to accommodate individual preferences. Accordingly, it is necessary to provide different types of units that residents can choose from or increased flexibility in the design of housing units or complexes to accommodate the diversity in preferences. Alternatively, given that the level and type of impairments may change as people age, a universal design that meets different requirements could best support ageing-in-place and thereby enhance QoL.

4.2. Design Consideration by Themes

4.2.1. Spatial Organization Surrounding the Sitting Area and the Sequential Space (Interior/Exterior)

There were different preferences in spaces that participants liked to stay; however, most participants liked to stay in their sitting space in their lounge. In the design of the space for people with restricted mobility, there should be careful consideration of the micro environment surrounding their sitting area (Figure 1). In particular, the sequence of space from the sitting area to the outside must be designed for access and control. The spatial organization should allow the layout of a chair (typically an adjustable armchair) and adjacent tables and/or shelves to ensure things are within the occupant's reach to enhance control of their environment. Consideration of preferred activities can ensure that the space can accommodate intended use. For example, given that watching TV is a common activity, layouts should permit location and proximity of TV options with respect to armchair location and in addition, the adjacency of any windows to avoid glare on the screen. There should be enough space for visitors in the quasi-public areas of the unit and a separation of the lounge from the bedroom. The spatial organization that allows residents to view visitors coming while the resident is seated improves their sense of control. The front door should be within sight of the sitting space as well as close enough for the voice to reach through the door. Windows should also be positioned to provide the resident outside views, but limit views from the outside to the inside.

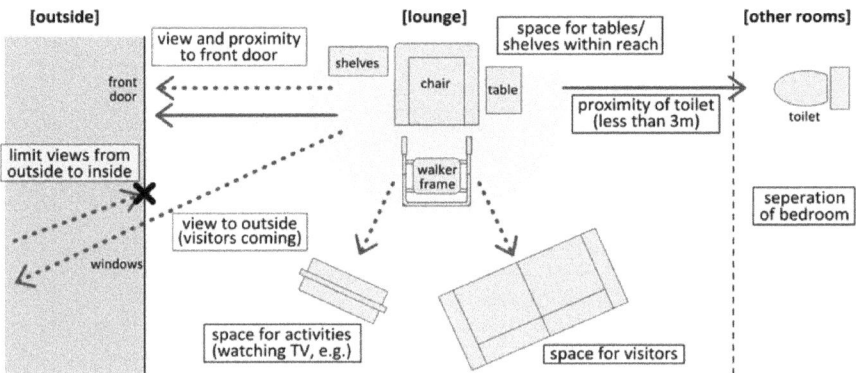

Figure 1. Diagram for spatial organization surrounding the sitting area.

Incontinence is a common problem for elderly people, the concerns of which can be worsened by restricted mobility. Locating a toilet as close to the sitting space or the bed as possible (less than 3 m) could address this issue for some people. However, for those with severe mobility concerns, accommodation should be made for toileting to occur in the lounge as well as in the bedroom through the use of a commode, or other devices. There should be enough consideration in the spatial organization of exterior space and placement of windows to meet the conflicting needs for high levels of privacy and other desires such as looking outdoors, welcoming visitors, or just enjoying the sunshine.

4.2.2. Storage

Consideration in the design of storage spaces with regard to interior elevation as well as necessary floor area is required for the common amount of objects and furniture. Built-in shelves of appropriate height are generally preferred particularly in the bathroom and the kitchen. The kitchens observed in this study generally had cupboards/shelves that were too high for ease of access by their intended user. In an attempt to provide enough storage in the limited space, often the storage was unusable for those with limited mobility. The kitchen should be redesigned or enlarged so that enough useable storage is provided.

4.2.3. Floor

Strategies for floor design with no level difference indoors as well as at the external door is required to meet the requirements of those with the highest levels of impairment. Interior floor design with no threshold could be a solution. There should also be consideration in the flooring materials with respect to maintenance, as people have a higher propensity to dirty the floor and a lower ability to clean it as the level of impairment increases. One resident wanted a non-slip tile floor for the entire unit, which could be easily cleaned by a steam cleaner; however, there is a common preference for carpet for warmth. There should be consideration for easily cleanable materials that are warm to the touch.

4.2.4. Door

Hinged doors, particularly when combined with a threshold with a level difference, are difficult for those walking with aids such as a trolley or a walker frame to manipulate. Sliding doors, that do not require much strength to open, are more suitable. The door serves to maintain privacy and to retain heat; however, they can be difficult to negotiate for those with limited mobility and can take up valuable space. For example, doors between the laundry and the bathroom, or the kitchen and the lounge could be removed.

4.2.5. Shower Area

Special consideration of the type of shower enclosure and the degree of fixture and flexibility of the shower head is required to enhance the independence of elderly people with mobility impairments. There should be no change of level in shower areas. For assisted showering, there should be enough space for drying as well as washing to accommodate both the resident and a caregiver. In addition, showering areas should have fittings that enable assistance out-of-sight of residents to maintain their privacy. The design of walls and fixtures that could be held by the elderly with both hands to support their balance increases their safety and thereby their independence and privacy.

4.2.6. Communal Space

In the design of communal space, there should be spaces that accommodate residents' preferred approach to socializing, such as meeting visitors in private common spaces as well as open organized activities. Flexible space and appropriate facilities should be provided to facilitate various preferred uses. There should be consideration in the accessibility and distance between the communal space and the unit to suit those with limited mobility and those using mobility aids. In addition, the location of toilets should be designed to meet the needs of those with incontinence.

4.2.7. Exterior

In the design of a complex, external pathways should not be close enough for passers-by to look inside compromising a resident's privacy. They also should be designed so that it can prevent the public from going through the housing site. The unit layout should take account of the aspect to the sun and the distance between units so as not to block the sun to another unit, particularly in the case of the tall buildings. The exterior space close to resident units, such as a porch, is used to accommodate various extra things, including mobility aids or other vehicles. The porch is also an important space for residents to enjoy the sunshine. The common space in close proximity to the units also has great potential to be used for interaction with other residents and accommodating guests. There should be enough consideration for the accessibility from the inside of the unit as well as the spatial layout that can accommodate chairs. In the design of the open space, there should be enough consideration for the layout and choice of trees with regard to resident's units, so that residents can fully enjoy the sunshine.

4.3. Design Strategies for Local-Authority Rental Housing to Accommodate the Elderly Occupants' Needs

Rental housing is typically an option for those who cannot afford to live in retirement villages. This is particularly the case for public-sector housing, where the eligibility criteria include levels of income and assets. The majority of units in the public-sector rental housing complexes for the elderly are bedsit or one-bedroom houses of 30–50 m^2. To improve the QoL while meeting the financial needs of the occupants, there should be special consideration in the design strategy, without significantly increasing the cost, which relates to the floor area as well as other architectural elements. Considering that a bedsit unit limits the opportunities to have guests in their house, the one-bedroom unit where the bedroom and the lounge are separated by a wall and a door should be made the standard. Using a sliding door between the two rooms, that can be kept both opened and closed, can increase flexibility in use and reduce the space taken for the swing of a hinged door, as well as being easier to open and close for elderly occupants with limited mobility. To provide the opportunity for guests to stay overnight, the lounge space should be designed so that it can accommodate at least a sofa-bed in addition to residents' own furniture requirements. Even with the limitation in floor area, there should be consideration in the bathroom design to accommodate two people in the space for assisted drying and showering, which may require more space than conventional bathroom designs. This does not necessarily mean greater cost—under a limited budget, it can be achievable with careful design of the whole house, including reducing unnecessary elements such as corridors and doors.

Common space can be also designed to accommodate objects and activities that cannot be accommodated in the dwellings. For example, mobility aids and other vehicles can be stored in the individual porch, or in a communal storage. In either case, there should be consideration for installing electric outlets for electric wheelchairs and mobility scooters as well as level access routes between such spaces and resident's units.

5. Limitation and Expansion of Research

Through this study of six high-needs elderly people, the implications for design have been identified; however, further research is required with more respondents to confirm and clarify these results. In addition, to explore the most suitable models of housing for the elderly, a greater analysis is required of the implications of ethnicity, gender, living arrangements, levels and types of impairments and type of housing, to explore the most suitable models of housing for the elderly. This paper includes limited consideration for cost aspects that can affect the design of public housing units, such as those argued by Leung et al. [30]. Findings of this paper include the micro-spatial use of the elderly in their house; however, in order to identify the optimal housing size, further investigation is required for their use of the space for possessions.

The design considerations derived from the investigation of spatial usage and the perception of the occupants will contribute to the improvement in the local-authority housing as well as other types of housing in the future. It can support the independent life of the high-needs elderly with greater QoL, and therefore provides one of the ways to overcome the demographic change in our society.

6. Conclusions

As the population ages, there will be an increased demand for housing that can accommodate those impairments that typically correspond with ageing. This paper examined the QoL of six elderly people with impairments living in local-authority rental housing. Analysis found that housing design has great potential to improve QoL of high-needs elderly residents in six aspects: Independence and activities; Sense of control; Privacy; Relationships; Quality of care; and Comfort. The design of housing that improves their QoL requires careful consideration for the micro-spatial organization surrounding the sitting area and sequential space towards the outside, to facilitate greater control and range of activities as well as providing adequate privacy and safety. There should be consideration of an expansion of space for accommodating the caregiver and facilitating meaningful relationships in the individual units and bathrooms, with careful design that will not significantly increase the cost. The design of indoor/outdoor common space should be flexible to accommodate comfortable relationships and activities, which is particularly effective in the design of complexes with small units. This paper also finds the design of housing that improves QoL requires solutions to accommodate a variety of conflicting needs derived from the diversity in user's preferences and the characteristics of their impairment. In the design of individual housing units and adjacent facilities, there is a greater need for reorganized or additional space to improve various aspects of QoL important for the high-needs elderly maintaining their autonomy and independence for as long as possible.

Author Contributions: All authors conceived and designed the research, Y.K. performed the data collection and analysis.

Conflicts of Interest: The authors declare no conflict of interest.

References

1. Hale, B.; Barrett, P.; Gauld, R. *The Age of Supported Independence: In Voices of In-Home Care*; Springer: New York, NY, USA, 2010.
2. Jaye, C.; Hale, B.; Butler, M.; McKechnie, R.; Robertson, L.; Simpson, J.; Tordoff, J.; Young, J. One of us: Stories from two New Zealand rest homes. *J. Aging Stud.* **2015**, *35*, 135–143. [CrossRef] [PubMed]
3. Statistics New Zealand. *2001 Census Population and Dwellings: Population Structure and Internal Migration*; Statistics New Zealand: Wellington, New Zealand, 2002.
4. Davey, J. "Ageing in Place": The Views of Older Homeowners on Maintenance, Renovation and Adaptation. *Soc. Policy J. N. Z.* **2006**, *27*, 128–141.
5. LiLACS NZ. *Intervals of Care Need: Need for Care and Support in Advanced Age*; University of Auckland: Auckland, New Zealand, 2017.
6. Ministry of Social Development. The New Zealand Positive Ageing Strategy. 2001. Available online: http://www.msd.govt.nz/ (accessed on 1 March 2018).
7. Greenbrook, S.R. *Village-People—The Changing role of Retirement Villages in New Zealand's Ageing Society*; School of Geography and Environmental Science, The University of Auckland: Auckland, New Zealand, 2005.
8. Johnson, A.; Howden-Chapman, P.; Eaqub, S. *A Stocktake of New Zealand's Housing*; New Zealand Government: Wellington, New Zealand, 2018.
9. Keeling, S. Later Life in Rental Housing. *Policy Q. Spec. Issue Ageing Popul.* **2014**, *10*, 54–59.
10. Statistics New Zealand. *2013 Census QuickStats about National Highlights—Tables*; Statistics New Zealand: Wellington, New Zealand, 2013.
11. Statistics New Zealand. *2013 Census QuickStats about People Aged 65 and over*; Statistics New Zealand: Wellington, New Zealand, 2013.
12. New Zealand Government. *Overview of the Government's Programme to Improve Social Housing in New Zealand*; Ministry of Social Development for the Social Housing Reform Programme: Wellington, New Zealand, 2015.

13. Kuboshima, Y.; McIntosh, J.; Thomas, G. Models of care and physical environments of current housing for the elderly: The possibilities of the rental housing for the dependent elderly in New Zealand. In Proceedings of the 33rd PLEA Conference, Edinburgh, UK, 2–5 July 2017.

14. Kuboshima, Y.; McIntosh, J.; Thomas, G. Care, Physical Environments and Dependency: The Design of Housing for the High Needs Elderly to Live Independently. In Proceedings of the AMPS CONFERENCE 10: Cities, Communities and Homes: Is the Urban Future Livable? Derby, UK, 22–23 June 2017.

15. Tester, S.; Hubbard, G.; Downs, M.; MacDonald, C.; Murphy, I. Frailty and institutional life. In *Growing Older: Quality of Life in Old Age: Extending Quality of Life*; Walker, A., Hennessy, C.H., Eds.; Open University Press: Maidenhead, UK, 2004.

16. Murphy, K.; Shea, E.O.; Cooney, A. Quality of life for older people living in long-stay settings in Ireland. *J. Clin. Nurs.* **2007**, *16*, 2167–2177. [CrossRef] [PubMed]

17. Nord, C. Architectural space as a moulding factor of care practices and resident privacy in assisted living. *Ageing Soc.* **2011**, *31*, 934–952. [CrossRef]

18. Chisholm, E.; Howden-Chapman, P.; Fougere, G. Renting in New Zealand: Perspectives from tenant advocates. *Kōtuitui: N. Z. J. Soc. Sci. Online* **2017**, *12*, 95–110. [CrossRef]

19. Bierre, S.; Howden-Chapman, P. *Telling Stories: The Role of Narratives in Rental Housing Policy Change in New Zealand*; Housing Studies: Sheffield, UK, 2017; pp. 1–21.

20. McIntosh, J.; Leah, A. Mapping housing for the disabled in New Zealand. *N. Z. Med. J.* **2017**, *130*, 69–78. [PubMed]

21. Higher Education Funding Council for England. *Guide to Post Occupancy Evaluation*; Higher Education Funding Council for England: Stoke Gifford, UK, 2006.

22. RIBA; Hay, R.; Bradbury, S.; Dixon, D.; Martindale, K.; Samuel, F.; Tait, A. *Building Knowledge: Pathways to Post Occupancy Evaluation*; Value of Architects, University of Reading, RIBA: London, UK, 2017.

23. Orihuela, P.; Orihuela, J. Needs, Values and Post-occupancy Evaluation of Housing Project Customers: A Pragmatic View. *Procedia Eng.* **2014**, *85*, 412–419. [CrossRef]

24. Pretlove, S.; Kade, S. Post occupancy evaluation of social housing designed and built to Code for Sustainable Homes levels 3, 4 and 5. *Energy Build.* **2016**, *110*, 120–134. [CrossRef]

25. Gupta, R. Embedding Post-Occupancy Evaluation into Architectural Education: From Specialism to Mainstream. In Proceedings of the 2nd Annual AAE Conference 2014: Living and Learning, Sheffield, UK, 3–5 September 2014; Parnell, R., Rajendran, L.P., Mahdizadeh, S., Eds.; School of Architecture, The University of Sheffield: Sheffield, UK, 2014; pp. 145–150.

26. RIBA Practice Team; Morris, N. Calls on Government to Commit to POE. 2017. Available online: https://www.architecture.com/ (accessed on 19 March 2018).

27. Ministry of Education. Post Occupancy Evaluations (POEs) of School Building Projects. 2017. Available online: https://www.education.govt.nz/school/property/state-schools/design-standards/post-occupancy-evaluations-of-school-building-projects/ (accessed on 29 March 2018).

28. Hiromoto, J. Post Occupancy Evaluation Survey Report. 2015. Available online: https://www.som.com/ideas/research/post_occupancy_evaluation_survey_report (accessed on 29 March 2018).

29. Lucas, R. *Research Methods for Architecture*; Laurence King Publishing: London, UK, 2016.

30. Leung, C.K.Y.; Sarpça, S.; Yilmaz, K. Public housing units vs. housing vouchers: Accessibility, local public goods, and welfare. *J. Hous. Econ.* **2012**, *21*, 310–321. [CrossRef]

© 2018 by the authors. Licensee MDPI, Basel, Switzerland. This article is an open access article distributed under the terms and conditions of the Creative Commons Attribution (CC BY) license (http://creativecommons.org/licenses/by/4.0/).

Article

Translating across Disciplines: On Coding Interior Architecture Theory to Advance Complex Indoor Environment Quality

Linda Pearce †

School of Art, Architecture and Design, University of South Australia, Adelaide, SA 5001, Australia;
linda.pearce@unisa.edu.au; Tel.: +61-8-8302-0202
† This text has been expanded from an original conference paper: Pearce, L.M. Sensory pleasure of interiority:
  Finding transdisciplinary research language for complex indoor environment quality. In Proceedings of the
  51st International Conference of the Architectural Science Association, Wellington, New Zealand, 29
  November–2 December 2017.

Received: 29 March 2018; Accepted: 14 June 2018; Published: 21 June 2018

Abstract: While indoor environment quality (IEQ) measurement is an established process, it omits the pleasure of interior environments, possibly due to its perceived subjectivity in the context of objective productivity and profitability. Given the significant commercial interior design industry, which engages with the complexity of indoor habitation, there exists an opportunity to expand the scope of IEQ appraisal through inclusion of the interior architecture discipline as an IEQ stakeholder. This theoretical paper reframes existing building appraisal as convergent methods that are contingent on the discipline and audience, and proposes a sequential mixed methods research process that allows subjective and objective research methods integration. Drawing on the interior architecture discipline, and its holistic 'interiority', a content analysis of selected theoretical texts identifies candidate quality components for future development and use in environment quality measurement. The intention of this process is to translate across the interior architecture and architectural science disciplines by coding interior architecture perspectives into possible measurable variables. These broader candidate variables would likely be more inclusive of the lived experience and agency of occupants of interior spaces. Furthermore, they offer the possibility for extended complex indoor environment quality data for future use in advanced statistics.

Keywords: interior design; interior architecture; indoor environment quality; methodology; convergent methodologies; human factors

1. Introduction

Buildings have interiors and interior designers often design those interiors. In solving the wicked problem of Green Building, the first part of this statement is being addressed through significant building science and architectural science research efforts together with the development and application of practical ratings tools, such as GreenStar, Leadership in Energy and Environmental Design (LEED), Building Research Establishment Environmental Assessment Method (BREEAM), and the WELL Building Standard. The second part of this statement, that interior designers have a significant role in interiors, needs further investigation in the context of aspirational human-oriented design solutions in Green Building. This discursive paper takes the proposition that the profession of interior designers and their scholarly discipline of interior architecture may have a unique and useful perspective for indoor environment quality (IEQ) and initiates the translation of these perspectives to IEQ appraisal.

Interior design emerged as a professional in the nineteenth century [1]. Debate about the delineations of interior practice and origins is ongoing [2,3], with interior design, also known as interior architecture and spatial design, and practitioners known as interior designers. Interior designers are influential in creating enclosed built environment spaces. As a profession, interior designers make up 31% of the built environment design professionals (excluding urban planners) in Australia (up from 28% in 2011, [4]). While not all interior design is done by interior designers, interior designers are responsible for the creative design and detailing of new buildings and renovated interiors that include, but are not limited to, spatial planning, fixtures, finishes, furniture, materials, and lighting. In addition to designing the functionality, aesthetics, and atmosphere of enclosed spaces, in commercial projects they interpret the commercial brand and organisational design, and translating these to a creative spatial design which maximises productivity ([1] (p. 71), [5,6]).

The interior design profession also contributes to vernacular trends. In Australian residential buildings, nearly as much is spent on furniture, floor coverings, and houseware goods as is spent on clothing ($1.2bn vs. $1.4bn, December 2016, ABS report 8501.0, [7]). Interiors are more than shelter, yet pleasure in all building classes, not just residential, is considered equal to productivity.

This paper is positioned in the broader definition of interior architecture, i.e., ' ... the design of structurally created interiors ... ' [1] (p. 2), which includes interior design, decoration, and an understanding of structure and services. The associated body of scholarly knowledge is referred to here as interior architecture. The broadness of the practice of this discipline is seen as both opportunity and challenge for inclusion in built environment quality [8]. Interior architecture scholars tend towards theoretical knowledge in arguing their history, design interpretation, and professional practice. For this knowledge to be of interest to a new audience, it needs to be translated.

This paper develops a recent conference paper [9] to open this translation process. It starts by noting that architectural science and building science have a history of inter-disciplinarity and convergent methodologies and provides precedents where other subjective theories have been coded for use in appraising indoor environment quality. This translation process is put into a methodological research context in Section 3. In Section 4, it applies this extended research process to translating interior architecture theory for use in architectural science research and provides a visual demonstration using a Green Building. Section 5 discusses how this theoretical knowledge and process can be further developed.

This paper should be read in the context of the *Special Issue: Human Factors in Green Building* which has called for discourse, as well as empirical research. This paper aims to provide a scholarly background to expand indoor environment quality, using a specific body of knowledge: interior architecture. It is not intended as a practical addendum to existing methods, but as a foundation for future scholarly research and practical application to new and expanded methods of IEQ appraisal.

2. Coding Precedents of Subjective Experience

IEQ has a history of adapting its data collection processes to include subjective measures and make it useful to researchers and practitioners. This section highlights the range of processes and some historical precedents of the methods of coding the subjective experience.

2.1. Implicit Adaptability of IEQ

Indoor environment quality ranges from a precise definition of thermal, acoustic, visual, and air quality measures [10] through to wider interpretations that include other human factors, such as control and size of space [11]. IEQ is also sometimes conflated with post occupancy evaluation (POE) [12] (Table 14.1, p. 172). While POE has historically offered flexible options for appraising a building [13], IEQ has also been coded and commercialised for use during the design stage, as indicated by the commercial sustainability of LEED, BREEAM, and GreenStar ratings tools [14].

Large-N post occupancy evaluations with IEQ successfully exploit the repeatability of survey test instruments that code specific IEQ components [15]. In contrast, small-N building studies use other data collection procedures, such as walkthroughs or interviews [16] to develop rich case studies. These methods are common to other disciplines: environmental psychology [17], environmental behaviour [18], design [19,20], or other interested research from outside of the built environment professions [21,22]. All of these use either self-reported measures or expert interpretation, or a combination of both, and many relate back to some physical measure of the corresponding building [12]. Others have approached IEQ medically, such as a complex stressor on occupants' physiology [23] or by using neurobehavioural tests in controlled settings [24].

When studies report interior design components in IEQ/POE studies, the descriptions can be brief. For example, in one study, 'office layout design' and 'look and feel' is all that is provided to describe the interior architecture [10]. There is also a variety of terms used. Indoor environment quality is the predominant term, with others using 'internal environment conditions' [25]. When discussed in interior architecture literature IEQ has been described as 'interior environment quality' [8]. These examples show that while there is a need to acknowledge the quality of interiors, the literature is inconsistent in processes and definitions. It is also flexible and suggests exploration and innovation.

2.2. Coding the Subjective—Precedents in Building Science

In IEQ, qualitative human attitudes and perceptions of the built environment have been translated into repeatable questions and efficient surveys. The coding process depends on the research discipline and audience, but the recognition that occupants are important is consistent in the three precedents summarised here: thermal comfort, POE, and light preferences.

Early last century, Bedford clearly described how he coded the responses from participant interviews in 1936 to create his sensation of warmth scale [26]. He also reflected on the usefulness of coding his interviews as a scale, concluding that the process is beneficial for his purpose:

> The use of an arbitrary scale cannot be avoided, but it may be thought that a more reasonable scale could be constructed by assuming a normal distribution of the personal feeling of warmth. This point has been examined, but it is found that the use of a scale based on this assumption does not significantly affect any of the conclusions set out in this Report. It has, therefore, been thought desirable to use the simple scale set out above. [26] (p. 19)

The ASHRAE thermal sensation scale presents a similar coding and standardisation of subjective experience [27,28] and is implicitly accepted (by its intended audiences) as a complementary test to other physical methods used to interpret thermal comfort for specialist [29] (p. 12 in Chapter 9) and generalist audiences [30] (pp. 158–178). Thermal comfort also extends to biological reward of sensory pleasure, alliesthesia [31,32] and combined with other perception codes, including, but not limited to, personalisation, control, furniture comfort, collaboration space, and other traditional IEQ [33].

In post-occupancy evaluation, while interviews and walkthroughs provide rich understanding, they are labour-intensive to both collect and interpret. Scale questions, often in detached Likert response format [34], code selected occupant perspectives and create benchmarking opportunities (e.g., [35]), but can also be designed for specific circumstances according to researcher interest, such as the 'friendliness' of classrooms [36].

As an alternative to Likert scales, semantic scales may be used to test extremes between two adjective pairs on a scale. These responses can then be used to determine underlying meaning constructs using factor analysis [37]. These methods are found in earlier environmental psychology examples investigating perceptions of light [38], or affective states in different interior environments [28,39]. These semantic scale examples are highly controlled environments, as is appropriate to the psychology protocols, but, in the latter case, the 'décor' variables are extremely limited: white vs. dark (a walnut panel) vs. blue walls vs. orange walls. From the perspective of design application, these parameters are not particularly useful due to the vague description and the changing design fashions. Furthermore, the semantic pairs used were collected by testing undergraduate students [40,41] and may not fully describe other occupants' experiences, or the interest of professional designers.

There is a clear need to collect subjective data efficiently using some form of repeatable test instrument, as has been demonstrated in thermal comfort, post-occupancy evaluation and environmental psychology. These are examples of where disciplines have responded to discipline needs; however, there are limits to the usefulness of the analysis when translating across disciplines from, say, IEQ for facilities management to useful inputs to interior designers.

3. Research across Disciplines

From one architectural science perspective, architecture is split, somewhat neatly, into art and science [42] (p. ix). A common critique of research of human factors in buildings is that it is subjective. Despite this, researchers and practitioners still attempt to integrate the subjective into architectural and building science, suggesting it is a necessary component of the topic. This section makes explicit one model of the integration process as a framework for integrating theoretical interiors knowledge.

Disciplines have their own taken for granted rules and scope of interest. Crossing these boundaries has been described as multi-disciplinarity, inter-disciplinarity, and trans-disciplinarity research [43] (p. 21). Others suggest that this is not helpful given that, while disciplines exist as separate specialisations, they are constantly evolving over time, making it difficult to consistently apply these terms [44].

Another approach is to consider the knowledge production context. It has been proposed that there are two 'modes' of knowledge production. Mode 1 refers to discipline-dependent scientific research processes used by independent scientists within an academic institution, whereas Mode 2 knowledge is 'socially distributed knowledge' created within a range of contexts with quality measured by its contextual value [45]. In the case of human factors in Green Buildings, the contexts of the coding precedents above suggest that indoor environment quality and post occupancy evaluation, by virtue of the range of practitioners, in both research and professional practice contexts, should be located as Mode 2 knowledge production. The application of this mixing of discipline knowledge can be described as convergent methodologies. In life and physical sciences, this is offered as a means of addressing complex real-world problems that have interconnected physical and social components and require a network of discipline expertise, and their specific tools, to solve the relevant query [46].

The term 'convergent methodologies' is also found in architectural science, but in the context of mixed-methods common to social science, where 'triangulation' is used as a metaphor used to integrate the findings [47]. Social science texts provide further instruction through reconciliation, or 'meta-inference', of parallel research strands [48]. Thus, rather than a network of experts, in architectural science the convergence is oriented towards networks of methods.

There seems to be two options for convergence. To include interior architecture knowledge into building science both convergence approaches need to be realised. First, as a mode 2 knowledge production process, interior architecture needs to be recognised as part of the network of expertise. Second, this expertise needs to be accessible and one approach (and there are others) is to translate it for use in building science. To facilitate this latter convergence, it is proposed to decouple methodology from method under the knowledge claim of pragmatism, and then demonstrate convergence as a sequential research continuum from recognising a surprising phenomenon to inclusion in research.

3.1. Decoupling Methodology and Method

Research methodology, how research is designed and the research methods deployed to answer research questions, and the quality of those answers, depends on the worldview of the researcher and their discipline [48]. Research quality adjudication differs between methodologies, depending on positivist or constructivist, objective or subjective, positions with disciplines using specific methodologies and taken for granted protocols [49] (p. 81). This complies with mode 1 knowledge production that uses strict discipline-specific protocols. Yet, in practice, as evidenced by IEQ and POE, this is clearly not the case and may be accidently innovative.

Separating methods according to knowledge claim has been queried. It is recognised that there is power contained within mixed-methods for interdisciplinary research:

... it is highly likely that much can be learned about generative and thoughtful mixed methods practice from the extraordinary explosion of provocative mixed methods empirical work and from more concerted *and* deliberate conversations across disciplines and fields of applied inquiry practice. [emphasis in original] [50]

This does not mean that anyone and everyone may create good-quality knowledge; rather, in the case of the built environment, inquiry should aim to develop 'informed judgement' to create 'responsive cohesion' within the built environment [51] (pp. 85–88). This is evident in post occupancy evaluation research precedents using both qualitative and quantitative data (e.g., [52,53]), suggesting implicit mixed methods.

Returning to methodology and worldviews, the location of mixed methods in the epistemological debate varies from constructivist according to architectural research methods [49] (pp. 218–219), to being technically independent of epistemology [54], to pragmatist [55], thus offering a symptom of how knowledge paradigms are continuously under debate [56].

Pragmatism offers an explanation about researching across disciplines for three reasons. First, it acknowledges the flexibility and continuous improvement needed in research methods. Pragmatism is described as a 'living philosophy' [57] (p. 4) where, rather than relying on expert beliefs, the perception exists that knowledge is 'fallible' and must be constantly refuted, or strengthened, to resolve 'doubt' as more evidence appears, through continuous evaluation [57] (pp. 15–19).

Second, it acknowledges that research is done for specific audiences and, when presenting knowledge, the intended audience must be convinced [58]. Where knowledge is found to be incorrect by the intended audience it could be rejected outright, but this is in danger of throwing out the knowledge baby with the fallible bathwater. 'Perspective fallibilism' allows knowledge to be considered as truth from a particular perspective but acknowledging the contradiction with another similar body of knowledge [57] (pp. 49–50), or may open up an interdisciplinary 'dialogical encounter' [59]. In a built environment performance evaluation, including IEQ, this is useful to consider where the intended audience consists of a wide range of stakeholders.

Third, it acknowledges the necessary junction between professional practice problem solving and scholarly knowledge creation. Pragmatism is attractive because it allows the inclusion of real-world practical knowledge, or praxis [60], and this makes it particularly useful in a practice-based academic discipline such as architecture [61]. Mixing of methods has been recommended for applied disciplines, including architecture research [61], thermal comfort investigations [62] using a mix of observational, survey, or other data (e.g., [63,64]), in architectural practice [65–67] and POE IEQ.

The purpose of the above discussion was to acknowledge current building and architectural research activities as being implicitly mixed and note that these do not fit neatly into the epistemology-methodology-method relationships that are described in research education texts [49]. This offers a freedom to seek new interpretations of the architectural science and building science research process that are inherently Mode 2 practical research and might include interior architecture in the network of expertise.

3.2. An Argument for Sequential Convergence of Research Methods

Extending architectural science beyond physics to include people in the research is not new. Last century, Hillier and Leaman raised limitations with scientific method, and discussed a number of paradoxes associated with the application of 'scientific certainty' to psychology and variability in human behaviour. As an alternative to physical 'spatial space' in a 'man-environment paradigm', they suggest a 'logical space' constructed by society and analogous Levi-Strauss' structural sociology, where social structures both describe and act on a population [68].

Gidden's later sociology theory of structuration considers that structure is created through recursive social practices includes social structure, but also acknowledges agency of individuals within social structure. Society should be studied:

... neither the experience of the individual actor, nor the existence of any form of societal totality, but social practices ordered across space and time. Human social activities, like some self-reproducing items in nature, are recursive. ... they are not brought into being by social actors but continually recreated by them via the very means whereby they express themselves *as* actors. In and through their activities agents reproduce the conditions that make these activities possible. [69] (p. 2)

In later work, Hillier [70] expressed concerns with applying Giddens' arguments, since Giddens is specific in his insistence that space is a social construct, whereas Hillier maintains space is a unique spatial paradigm describable separately as space syntax. Taking the broader position discussed here, the spatial paradigm, and its representations, could also be considered as part of a larger recursive mixed methods discussion—providing rich visualisation in and of itself—yet also contributing to discussions about social production of space in the context of different audiences, which is a result of power structures and other privilege.

Returning to Giddens, structuration is of interest to designers since it reminds designers to acknowledge that their designs are contingent on occupation and time [71]. Applying this to building appraisal, rather than rejecting the scientific empiricism of environmental space for a logical space, this opens up the re-examination of positivist approaches towards building appraisal as both providing limits to occupation and a response to occupant agency. This opens up the possibility of surprising occupation, as described by interior architecture, and learning from it.

In science and technology studies (STS), Latour argues that the separation of science from non-science, objectivity from subjectivity, never existed, and that this separation is a constructed political decision, which should be reversed [72] (p. 144).

Half of our politics is constructed in science and technology. The other half of Nature is constructed in societies. Let us patch the two back together, and the political task can begin again. [72] (p. 144)

Similarly, philosopher A.N. Whitehead argues that we should not 'bifurcate' nature because there is an interaction between cause of awareness and awareness:

... everything perceived is in nature. We may not pick and choose. For us the red glow of the sunset should be as much part of nature as are the molecules and electric waves by which men of science would explain the phenomenon ... (Whitehead CN29 in [73]) (p. 33)

This is particularly useful since it acknowledges that different interpretations of the world, objective physics and subjective beauty, exist simultaneously, implying that we naturally use different methods to understand our world. This does not mean that the methods are wrong (heat transfer physics is clearly useful), but stepping outside of a specific community opens up choices about research methodology and methods.

Environmental psychology is an obvious gateway to user experience within built environment research [74]; however, ambiguous yet persistent experience of interior is not necessarily covered to suit the building *design* community to put into practice. While there are research efficiencies and validities associated with psychology's science methods, if variable selection is undertaken without designer input, this reductionist approach of coding or quantising indoors is of limited use.

The interior architecture scholar must consider the opportunities: their knowledge base includes ephemerality that may not be knowable beyond interior theorists and personal narrative, the latter clearly important, as seen by the commercial success of the building adaptation industry, but must also engage with scientific methods and inter-disciplinarity. This is where pragmatism and reframing the objective/subjective paradigm is useful for researching across disciplines.

Decoupling methods and methodology and mixing methods under pragmatism epistemology, considers positivist and constructivist research methods as complementary and inter-related through abductive logic [49] (pp. 34–35), [48] (p. 89). Abductive logic argues that knowledge starts with observing a surprising phenomenon, initiating a circular deduction and induction knowledge creation process (Figure 1a). This is a sequential mixing of research methods where triangulation is a convergent dialogue between theoretical statements and empirical observations [75].

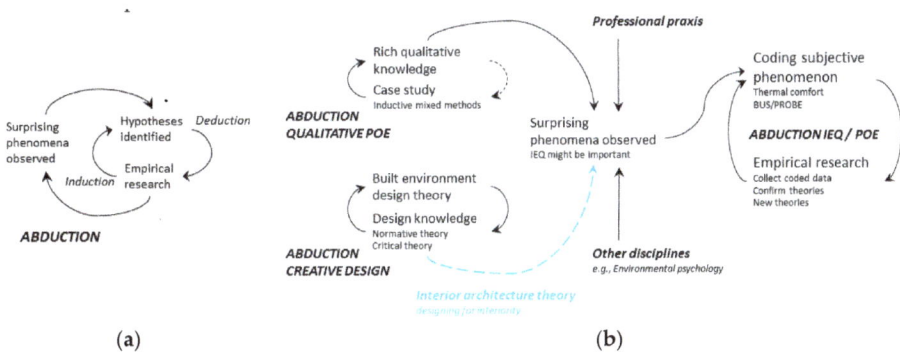

(a)   (b)

Figure 1. (a) Abduction; and (b) interconnected/sequential abduction in IEQ/POE.

Indoor environment quality measures come from somewhere (Figure 1b). Architectural science clearly uses surprising phenomena from professional practice and research to trigger new lines of inquiry (e.g., [52]). Someone observed an effect or a need and developed *useful* hypotheses and tools to quantify indoor environment quality. This is abduction in practice. The start of this process is coloured by the originator's tacit knowledge. Any extension of professional praxis [76] (pp. 37–44) will influence the process, but will also provide 'practical wisdom' [60]. Current IEQ approaches are fit for (the current) purpose. This paper is interested in expanding the existing inductive origins of environment quality. The inclusion of interior architecture sources is just a sequential continuum of the abduction process.

4. Interior Architecture as Source Discipline

This section reviews briefly the interior architecture discipline and then interrogates selected literature for new interpretation of interior occupancy as a source of observed surprising phenomena. This process is shown diagrammatically in Figure 2.

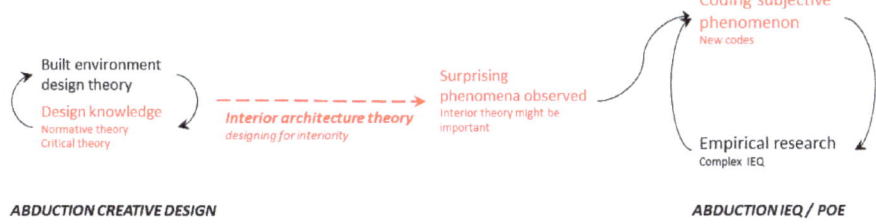

Figure 2. Pathway for introducing interior architecture theory into IEQ, based on Figure 1.

4.1. Interior Architecture—A Brief View from the Theoretical Scholars

Interior practitioners draw on rich theoretical traditions. Interior Architecture (IA) is ' . . . the design of space through human occupation' [77] (p. 8), or ' . . . design of the near environment' [8], such that:

Designers act upon interiors through multiple entry points that include atmospheric conditions like color and light, understanding the client's needs, giving form and shape to materials, and unifying these elements into a captivating design. [78] (p. 11)

Thus, it is taken for granted that occupants should feel pleasure about the built design. Occupancy is seen as 'familiarity' with an interior due to ' . . . the close proximity of people, objects, and space' [77] (p. 3). It is also concerned with the temporality of designed spaces, particularly the need to adjust infrastructure to occupants as their needs change [2,74], leading to shorter lifecycles for interior architecture than architecture.

The interior architecture knowledge base comes from a wide range of research methods and knowledge claims, such as, but not limited to, theoretical approaches [79], including envronmental psychology theory [20]; empirical approaches, such as controlled experimentation, e.g., [39], historical, e.g., [80], and social sciences, e.g., [81]; and practiced-based research by design [82].

Interior preferences are highly complex and subjective, and are both individual and constructed through socialisation [83]. In professional design, aesthetics and pleasure of occupancy of a space are commensurate with its functionality objectives. While the term 'space' is often used in describing an interior (e.g., [1] (pp. 114–143)), the term 'interiority' offers an active view of occupying space:

Interiority is that abstract quality that *enables the recognition and definition of an interior*. It is a theoretical and immaterial set of coincidences and variables from which "interior" is made possible. [84] (p. 112, emphasis added)

Interiority is development of enclosure and boundaries, originating from historical interior design [85], but also the pure sensory engagement of linking personal with spatial interiority [86] (p. ix). It is a developing concept, beginning as a social theory and moving to a recursive activity of spatial construction [87].

This theoretical literature of interior architecture offers a significant body of knowledge about the holistic understanding of interiors. This offers additional qualitative assessment to architectural and building science in the form of alternative phenomenology and interpretations; however, translating theory into coded formats needs deliberation. It has been asserted that itemising spatial components, and breaking down interiority, is not particularly useful for interior architecture, and it is recommended to:

. . . recognise that multiple paradigms operate simultaneously—the sensorial experience, the cognitive or thoughtful, evaluative experience, and the immediate confrontation or immersive experience—[so that] a more holistic understanding is facilitated. [88]

In contrast, the duality of environmental quality components is also noted: light can be measured scientifically and holistically as an 'antediluvian affect' [89]. Similarly, from the environment psychology perspective, it is acknowledged that a range of checklists and test instruments are needed to fully cover interior features as well as perspectives [20]. Thus, measuring environment quality is not an either/or situation: both quantitative and qualitative, objective, and subjective methods have their individual validity and purpose, reinforcing each other, suggesting that interior architecture is not hostile to scientific methods.

The research efficiencies of quantitative surveys used in architectural and building science remain attractive, particularly for generating large datasets. The following speculative exercise tests the coding processing using interior architecture as a new source of variables for IEQ.

4.2. Code Development from Interior Architecture

Following the sequential pragmatic abductive process in Figure 2, this section searches interior architecture theory for new codes for future inclusion in IEQ. It does this by using content analysis [90] (pp. 282–285) in which selected interior architecture texts are examined for new words and phrases to describe interiors. The interior architecture texts were selected because they are used in interior architecture education at the author's university workplace and represent contemporary thought about the interior architecture discipline. The source authors are listed and the edited books are coded as B + W = Brooker and Weinthal [77]; W = Weinthal [78]. The terms presented are those that provide 'surprising phenomena' and expanded interpretations relative to current IEQ.

Table 1 lists preliminary sense variables found in these texts. They provide nuanced interpretations of interiors and acknowledge that interiors change over time. While personal data collection is appropriate for some of these, visual data can be used to make observations of the application of these codes. This is illustrated here with images of Level 5 of the Jeffery Smart Building at the University of South Australia (Adelaide 35° S). This library and learning centre was designed by John Wardle Architects in association with Phillips Pilkington Architects. It opened in 2014 and was certified with a 5-Star Green Star—Education Design (v1) rating in 2016 [91].

The photos in Figure 3 show that there are distinct zones as indicated by the interior architecture's furniture and fittings selections. This creates near vs. far vision within the space. These are also differentiated be rectilinear and regular forms (library stacks), technology-rich areas (individual workstations), and the curved edges of the pendant lights and their relationship to the stand-up desk on a hard floor covering. In this case, locations could be characterised with codes, such as 'fixture edges' (rectilinear vs. curved, or strict vs. casual). There is also an 'internal visual distance' (near, mid, far).

Table 1. Selected preliminary content analysis of selected key texts—senses.

IA Topic	Source	Content
Senses—acoustic intimacy	Pallasmaa (W) Cantwell (Ch 38, B + W) von Drathen (W)	Presence/absence/time marker Harshness/softness/tranquillity Directionality
Sense—Sense of body/bodily resonance in space	Pallasmaa (W) Cantwell (Ch 38, B + W) von Drathen (W)	Scale/volume Interaction Gravity—apparent vs. defying
Senses—Vision extended (seen vs. potentially touched)	Pallasmaa (W)	Near vs. far Surfaces, contours, edges Agreeableness/unpleasantness Affection/indifference/stress
Senses—touch	Pallasmaa (W) Cantwell (Ch 38, B + W) von Drathen (W)	Texture and density Weight; Eye vs. body Temperature and light
Senses—olfactory	Pallasmaa (W) Parkinson (Ch 22, B + W)	Memory Association

(a)

(b)

Figure 3. Level 5 northern study area, Jeffrey Smart Building, University of South Australia: (**a**) view to the southwest; and (**b**) view to the north (photos copyright of the author).

Table 2 addresses the interpretation of the interior enclosure. This enclosure is not the building envelope familiar to Architectural Science. Rather, it provides an exemplar from the perspective of the Interior Architecture discipline in its focus on the experience of being in a space and looking out.

In the example, there is a distinct interior fashion in the furniture selections, materials and colours. There are two tables shown in Figure 3. The table on the right is at traditional office desk seating level. The one on the left is higher and, while it has high chairs around it, it is also useable when standing, making it contemporary with current design trends for standing workplaces. This change in height contributes to the design style, but also to near vs. far vision compared to the previous table.

There are two types of permeability in the space. The first is the traditional view out of the building. The views out of the space to the exterior are obstructed by automatic blinds (Figure 3). This is due to the time of day that the photos were taken (summer morning in January, so north and east side blinds are down).

Within the space, each zone is delineated by changes in fixtures, furniture, colours and materials. These signify interiority and this creates small interiors within a large interior. There is visual permeability between each space with changes in privacy and openness. This could be coded as 'interior permeability' (low vs. high). The wall seat joinery in Figure 4 is functional, yet it does not adhere to an 'instantly detectable function' or 'affordance' as described in environmental psychology [20] (p. 30). Its unusual design shape brings attention to the wall, where the perforations of the acoustic panelling provides texture. The window mullions in Figure 5 are angled. The photograph, taken on an angle, highlights the texture this design decision makes to the space. Both of these examples, when compared to a plain plasterboard wall or glazed curtain wall, are high 'wall texture'.

Table 2. Selected preliminary content analysis of selected key texts—interior enclosure.

IA Topic	Source	Content
Historical and Geographical design influence/Fashion trends	Massey (Ch1, B + W) Scott (Ch 10, B + W) Shyder (Ch 29, B + W) Sparke (Ch 39 B + W)	Design style/hybrid Diffusion Flexibility Fashion
Threshold/connection between public and private	Griffith Winton (Ch 3, B + W) Parkinson (Ch 22, B + W) Moreno (Ch 26, B + W) Verghese and Smith (Ch 36, B + W)	Entrance openness Sense of privacy Permeability in and out (views and physical)
Materials and colour and surfaces	Verghese and Smith (Ch 36, B + W) Bachelor (W) Weinthal (W) Seigel (W)	Texture and Moulding, Light Cultural norms of colour (national, commercial, fashion) Safety of materials
Technology	Keeble (Ch 37, B + W) McQuire (H)	Comfort (heat, light) control Surveillance/Linkage Domestic vs. industrial tech Work vs. pleasure technology Ambivalence vs. defined outcome Participation

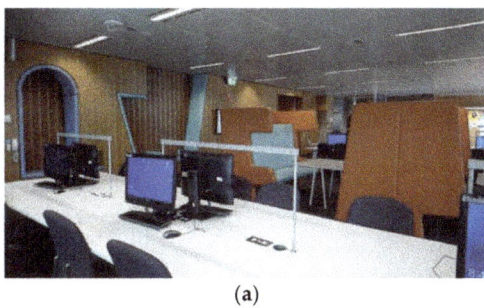

(a)   (b)

Figure 4. Level 5 southern study area, Jeffrey Smart Building, University of South Australia: (**a**) contextual view towards the southwest; and (**b**) wall chair detail (photos copyright of the author).

Occupied spaces generally include some form of spatial agency. Table 3 addresses the effects of agency on spaces and has grouped together variables associated with personalisation under material culture, where everyday objects, and their deliberate arrangement, contribute to meaning and occupation. Here, too, Interior Architecture offers a wider and more nuanced evaluation of interiors [92].

Table 3. Selected preliminary content analysis of selected key texts—material culture.

IA Topic	Source	Content
Material culture (fixtures, fittings, decoration, furniture, that dress an interior)	Griffith Winton (Ch 3, B + W) Massey (Ch 35, B + W) Blauvelt (W) Schouwenberg (W) Helguera (W) Betsky (W)	Functional/everyday objects Pleasure objects Exhibition/installation of objects Participatory action of design/decoration Observed/reported/Preference

The images were taken prior to the start of the university term, so there is little evidence of use and occupation agency through personal and moveable objects. The stationary and fixed everyday objects used are coordinated both in materials and colour and demonstrate deliberate interior design agency to create symbolic meaning of a contemporary learning space.

The example here demonstrates the code of 'exhibition' where the installation of objects is present. The selection of pendant lights of variable size is an installation of objects in space above the high table (Figure 3). While the task lighting could have been provided with recessed fluorescent lights, as is done elsewhere, the design selections here create a place within the space through this installation. The wingback chairs in both Figures 4 and 5 also represent deliberate decisions to select functional objects with a novel form that suggests an installation of an object rather than a functional seat. This is similar to environmental psychology's collative properties of a room and its 'surprisingness' [20] (p. 285). Here, the designers have added complexity and a gradual reveal of possibilities rather than an instant understanding. Different audiences will understand this differently: for example, student users will interpret this space differently to professional designers whose principal concern is the interiority and atmosphere, yet both audiences are correct.

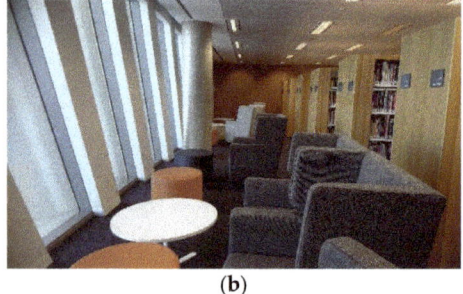

(a)   (b)

Figure 5. Level 5 southern study area, Jeffrey Smart Building, University of South Australia: (a) contextual view towards north east; and (b) west study area (photos copyright of the author).

Table 4 presents examples of the Interior Architecture discipline's approach in interpreting user experience. Noting that Interior Architecture draws from a wide range of disciplines, the influence of environmental psychology is evident; however, perspectives, such as immersion and engagement with spatial design, offer additional perspectives for review. This content is similar to codes currently used in IEQ. For these, survey test instruments are most appropriate for collecting user experiences. Longitudinal visual data could capture changes in interiors, such as temporary installations, and use anthropological research methods. Alternatively, the author is currently investigating mobile eye tracking technology combined with wearable technology to capture biological responses to represent the user experience.

Table 4. Selected preliminary content analysis of selected key texts—user experience.

IA Topic	Source	Content
Desire and delight	Moreno (Ch 26, B + W) Parkinson (Ch 22, B + W)	Immersion Preference/Liked
Transience/change	Farrelly (Ch 11, B + W) Littlefield (Ch 17, B + W) Moreno (Ch 26, B + W)	Preference Liked Permanent vs. temporary
Health and wellbeing through design for operational rationality vs. compassionate interior design	Parkinson (Ch 22, B + W)	Natural light, noise reduction, layout, views Engagement with spatial design Compassionate/welcoming space Emotional/physical stress Psychological/social support Overload/Peace/Stimulation Movement agency
Spirit of place/meaningful occupation	Farrelly (Ch 11, B + W) Verghese and Smith (Ch 36, B + W) Cantwell (Ch 38, B + W)	Likely a combination of other variables, e.g., factor analysis
Experience and familiarity of spatial environment	Verghese & Smith (Ch 36, B + W)	Time spent in environment Peripheral vs. primacy State of mind Associate physical/memory

The example here is a relatively new construction with less than four years of occupation. While it is classified as a 'Green Building', from the perspective of interior architecture, it may perform better than other buildings because the fabric is newer, has less deterioration, and is well maintained. Furthermore, it may have a better quality interior design because more design effort may have been expended on it, as is consistent of expectations for landmark buildings. The consequence of this is that the final configuration of a Green Buildings may be contingent on the recursive and socially constructed forms of its interior design. Thus, inclusion of the interior architecture discipline theory offers additional content and codes to incorporate into indoor environment quality appraisal. For the educational building, this would require additional terms to describe the indoor environment quality, such as fixture edges, internal visual distance, internal permeability, wall texture, and exhibition, in this case, with possibly more if user experience and longitudinal occupation observations are included.

5. Discussion and Conclusions

5.1. Complex Indoor Environment Quality

Architectural science and interior architecture use different words—indoor and interior, respectively. Where indoor environment quality is a set of physical measures with some preference variables, interior literature is broader, exploring the holistic phenomenological and interiority experience of the body. Through considering the sequential convergence of methods, it allows theory to influence empirical research. Interiority has been introduced here as a sense of enclosure, rather than as a physical enclosure. This intellectually frees up the reliance on building fabric, but it should not remove the building fabric from the research challenge: maintaining an exclusive position on either the scientific or humanist side does not progress interior architecture (or other building knowledge) and that, in practice, physics and interiority (phenomenological or other theory) are interconnected through human experience:

> Pallasmaa, Murcutt, and Zumthor's influence on designers has been broad but not received conditionally because of their unusual confidence in the dominant need of the body as it meets the forces of nature as the generator of architecture. [89]

Thus, any measure of environment quality needs to be explicit in its starting point and scope. Indoor environment quality measurements are achievable, but potentially limited, although the limitations may be valid depending on the intended concerned audience. This paper suggests that this scope could be extended and, using the interior architecture theoretical knowledge presented here, a more accurate name might be *interiority* environment quality. The risk with this is two-fold: first, the audience, architectural and building science, may not be fully aware of the breadth of interiority theory, and, second, though accurate, the use of another discipline-specific term may restrict future inter- and trans-disciplinarity.

An alternative term is *complex indoor environment quality*. This is indoor environment quality that is expanded to included variables that capture more of the complexity and richness of lived experience, including the pleasure of interior habitation, as described by interior architecture theory.

5.2. Future Development and Application

The next step is to refine the codes. This will include systematic searches of key peer-reviewed interior design journals to triangulate and confirm candidate code categories and descriptions, which can then be tested with pilot studies. This process will also need to be informed by the type of space and its intended use.

Currently, it is envisaged that there are three applications for an expanded IEQ variable set. The first application will expand indoor environment quality appraisal, possibly as fine detail to post-occupancy investigations (e.g., [93]). It is anticipated that there is a core variable group with additional clusters of variables based on interior spatial function, e.g., dependent variables, such as productivity and satisfaction, will vary from the residential to the workplace to other spatial classifications.

The second application aims to go beyond satisfaction and productivity and look for more complex relationships. This application aims to use codes in statistical analysis, such as inferential statistics, factor analysis, and structural equation modelling to represent the constructs underlying interior environments, all as a companion to other qualitative methods. This is returning to early environmental psychology semantic differential approaches [39–41], but applied in naturalistic settings, using variables relevant to interior architecture appraisal.

The third application is as input for 'big data' analytic methods and, in particular, network patterns and data visualisation [94]. This latter approach offers a paradigm change in representing interiors since, rather than focussing on cognitive interactions of body- and neuro-typical occupants, large datasets allow for visualisation of occupant diversity. It is anticipated that these datasets can be created through location data and biophysical data available from wearables. In interior architecture the relationship of pleasure in occupancy to, say, productivity, can be investigated alongside typical IEQ measures. These large data sources provide naturalistic data for parameter relationship testing, as well as sources for inductive research to test emerging relationships between occupant, interior environment, and technology.

5.3. Conclusions

While architectural science and building science provide objective knowledge through physical built environment appraisal methods, this paper speculated that there are opportunities to extend our knowledge of habitable space through integrating subjective knowledge from interior disciplines, known as interior architecture, interior design, and spatial design. This syncretic perspective is intended to advance convergent methodologies and methods for the purpose of better understanding the occupation of interior space.

Using existing post occupancy evaluation and indoor environment quality exemplars, this paper reiterated these as convergent methodologies and positioned them as mode 2 knowledge, in which complex contextual problems require a network of experts and disciplines for resolution. It then positioned their research under the pragmatism knowledge claim. This then allowed discussion about the mixing of methods without the restriction of positivist and constructivist knowledge claims, and used abductive logic to relate objective research to subject research as a continuum, thus removing discipline exclusivity. It was then argued that POE and IEQ variables used come from intelligent practical observation of surprising phenomena using sequential methods.

Design and occupation occurs within a social structure, but also with user agency. This position of structuration provides a framework for understanding the interior architecture discipline. The IA theoretical knowledge base offers rich phenomenological and sociological interpretation of the experience of 'interiority', the nuanced sense of being within a defined space that privileges pleasure of occupancy. This paper proposed that this abstract experience and language of interiority could provide additional human factor variables for exploration.

The abductive coding process was applied to the interior architecture discipline. Two interior architecture teaching texts were mined for new perspectives of IEQ and presented here as a preliminary content analysis. These were further demonstrated using visual examples of a tertiary education building which is acknowledged as a Green Building to demonstrate how the codes could be interpreted. The candidate codes require further investigation and validation for use in quantitative IEQ research, which is a complement to rich qualitative work. This theoretical paper presented the start of this process and will form the basis for future work to integrate holistic experience into new complex indoor environment quality appraisal through statistical and 'big data' quantitative methods.

Funding: This paper received no external funding.

Acknowledgments: Photographs of the Jeffery Smart Building are used with permission, and acknowledge the University of South Australia as the client and owner of the building, and John Wardle Architects in association with Phillips Pilkington Architects as the building architects. The author thanks Zhonghua Gou, editor of the Special Issue: Human Factors in Green Building for the opportunity to develop the conference paper, and the Interior Architecture program staff at the UniSA School of Art, Architecture and Design for the rich discussions.

Conflicts of Interest: The authors declare no conflict of interest.

References

1. Edwards, C. *Interior Design: A Critical Introduction*; Berg: Oxford, UK, 2011.
2. Brooker, G.; Stone, S. *Basics: Interior Architecture: Form and Structure in Interior Architecture*, 2nd ed.; Bloomsbury: London, UK, 2016.
3. Plunkett, D. The profession that dare not speak its name. In *The Handbook of Interior Architecture and Design*; Brooker, G., Weinthal, L., Eds.; Bloomsbury Academic: London, UK, 2013; pp. 93–104.
4. ABS. Australian Census OCCU Occupation Group 232 Architects, Designers, Planners and Surveyors 2011. 2016. Available online: http://www.abs.gov.au/websitedbs/D3310114.nsf/Home/2016%20TableBuilder (accessed on 8 January 2018).
5. Moreno, S. Designing desire. In *The Handbook of Interior Architecture and Design*; Brooker, G., Weinthal, L., Eds.; Bloomsbury Academic: London, UK, 2013; pp. 364–378.
6. Myerson, J. The evolution of workspace design: From the machine to the network. In *The Handbook of Interior Architecture and Design*; Brooker, G., Weinthal, L., Eds.; Bloomsbury Academic: London, UK, 2013; pp. 213–225.
7. ABS. *8501.0 Retail Trade, Australia: Table 12. Retail Turnover, State by Industry Subgroup, Seasonally Adjusted*, 3/3/17 ed.; Australian Bureau of Statistics, Government of Australia: Canberra, Australia, 2016.
8. Theodorson, J. Energy, daylighting, and a role for interiors. *J. Inter. Des.* **2014**, *39*, 37–56. [CrossRef]
9. Pearce, L.M. Sensory pleasure of interiority: Finding transdisciplinary research language for complex indoor environment quality. In *Back to the Future: The Next 50 Years, Proceedings of the 51st International Conference of the Architectural Science Association, Wellington, New Zealand, 29 November–2 December 2017*; The Architectural Science Association and Victoria University of Wellington: Wellington, New Zealand, 2017; pp. 573–582. Available online: http://anzasca.net/paper/sensory-pleasure-of-interiority-finding-transdisciplinary-research-language-for-complex-indoor-environment-quality/ (accessed on 14 June 2018).
10. Al Horr, Y.; Arif, M.; Kaushik, A.; Mazroei, A.; Katafygiotou, M.; Elsarrag, E. Occupant productivity and office indoor environment quality: A review of the literature. *Build. Environ.* **2016**, *105*, 369–389. [CrossRef]
11. Sakhare, V.V.; Ralegaonkar, R.V. Indoor environmental quality: Review of parameters and assessment models. *Archit. Sci. Rev.* **2014**, *57*, 147–154. [CrossRef]
12. Mallory-Hill, S.; Westlund, A. Evaluating the impact of green building on worker productivity and health: A literature review. In *Enhancing Building Performance*; Mallory-Hill, S., Preiser, W.F.E., Watson, C.G., Eds.; Wiley: Somerset, UK, 2012; pp. 167–178.
13. Preiser, W.F.E. *Building Evaluation*; Plenum Press: New York, NY, USA; London, UK, 1989.
14. Doan, D.T.; Ghaffarianhoseini, A.; Naismith, N.; Zhang, T.; Ghaffarianhoseini, A.; Tookey, J. A critical comparison of green building rating systems. *Build. Environ.* **2017**, *123*, 243–260. [CrossRef]
15. Baird, G.; Leaman, A.; Thompson, J. A comparison of the performance of sustainable buildings with conventional buildings from the point of view of the users. *Archit. Sci. Rev.* **2012**, *55*, 135–144. [CrossRef]
16. Newton, C.; Wilks, S.; Hes, D.; Aibinu, A.; Crawford, R.H.; Goodwin, K.; Jensen, C.; Chambers, D.; Chan, T.-K.; Aye, L. More than a survey: An interdisciplinary post-occupancy tracking of ber schools. *Archit. Sci. Rev.* **2012**, *55*, 196–205. [CrossRef]
17. Gifford, R. *Environmental Psychology: Principles and Practice*, 5th ed.; Optimal: Colville, WC, USA, 2014.
18. Wener, R.E.; McCunn, L.J.; Senick, J. Did that plan work? Post-occupancy evaluation. In *Research Methods for Environmental Psychology*; John Wiley & Sons, Ltd.: Hoboken, NJ, USA, 2016; pp. 249–269.
19. Zeisel, J. *Inquiry by Design: Tools for Environment-Behavior Research*; Cambridge University Press: Cambridge, UK, 1984.
20. Gosling, S.D.; Gifford, R.; McCunn, L.J. The selection, creation, and perception of interior spaces: An environment psychology approach. In *The Handbook of Interior Architecture and Design*; Brooker, G., Weinthal, L., Eds.; Bloomsbury Academic: London, UK, 2013; pp. 270–281.
21. Edgerton, E.; McKecknie, J.; McEwen, S. Building better schools: Methodological concerns and the need for evidence-based research. In *Architectural Quality in Planning and Design of Schools: Current Issues with Focus on Developing Countries*; Knapp, E., Noschis, K., Eds.; Colloquia: Lausanne, Switzerland, 2010; pp. 43–51.
22. Simon, N.S.; Evans, G.W.; Maxwell, L.E. Building quality, academic achievement and self-competency in new york city public schools. In *School Building Design and Learning Performance: With a Focus on Schools in Developing Countries*; Knapp, E., Noschis, K., Pasalar, C., Eds.; Colloquia: Lausanne, Switzerland, 2007.

23. Bluyssen, P.M.; Janssen, S.; van den Brink, L.H.; de Kluizenaar, Y. Assessment of wellbeing in an indoor office environment. *Build. Environ.* **2011**, *46*, 2632–2640. [CrossRef]

24. Lan, L.; Lian, Z. Use of neurobehavioral tests to evaluate the effects of indoor environment quality on productivity. *Build. Environ.* **2009**, *44*, 2208–2217. [CrossRef]

25. Menadue, V.; Soebarto, V.; Williamson, T. The effect of internal environmental quality on occupant satisfaction in commercial office buildings. *HVAC&R Res.* **2013**, *19*, 1051–1062.

26. Bedford, T. *The Warmth Factor in Comfort at Work: A Physiological Study of Heating and Ventilation*; Great Britain Industrial Health Research Board, His Majesty's Stationery Office: London, UK, 1936.

27. Rohles, F.H., Jr. The revised modal comfort envelope. *ASHRAE Trans.* **1973**, *79*, 52–59.

28. Rohles, F.H., Jr.; Milliken, G.A. A scaling procedure for environmental research. *Proc. Hum. Factors Soc. Annu. Meet.* **1981**, *25*, 472–476. [CrossRef]

29. ASHRAE. *Fundamentals: Si Edition; Chapter 9, Thermal Comfort*; American Society of Heating, Refrigerating and Air-Conditioning Engineers Inc.: Atlanta, GA, USA, 2013.

30. Roaf, S.; Crichton, D.; Nicol, F. *Adapting Buildings and Cities for Climate Change: A 21st Century Survival Guide*, 2nd ed.; Elsevier/Architectural Press: Oxford, UK, 2009.

31. De Dear, R.; Auliciems, A. Validation of the predicted mean vote model of thermal comfort in six australian field studies. *ASHRAE Trans.* **1985**, *HI-85-09*, 452–463.

32. De Dear, R. Revisiting an old hypothesis of human thermal perception: Alliesthesia. *Build. Res. Inf.* **2011**, *39*, 108–117. [CrossRef]

33. Candido, C.; Kim, J.; de Dear, R.; Thomas, L. Bossa: A multidimensional post-occupancy evaluation tool. *Build. Res. Inf.* **2016**, *44*, 214–228. [CrossRef]

34. Carifio, J.; Perla, R. Ten common misunderstandings, misconceptions, persistent myths and urban legends about likert scales and likert response formats and their antidotes. *J. Soc. Sci.* **2007**, *3*, 106–116. [CrossRef]

35. Baird, G.; Field, C. Thermal comfort conditions in sustainable buildings—Results of a worldwide survey of users' perceptions. *Renew. Energy* **2013**, *49*, 44–47. [CrossRef]

36. Douglas, D.; Gifford, R. Evaluation of the physical classroom by students and professors: A lens model approach. *Educ. Res.* **2001**, *43*, 295–309. [CrossRef]

37. Osgood, C.; Suci, G.; Tannenbaum, P. *The Measurement of Meaning*; University of Illinois Press: Urbana, IL, USA, 1957.

38. Flynn, J.E.; Spencer, T.J. The effects of light source color on user impression and satisfaction. *J. Illum. Eng. Soc.* **1977**, *6*, 167–179. [CrossRef]

39. Rohles, F.H.; Wells, W.V. Interior design, comfort and thermal sensitivity. *J. Inter. Des.* **1976**, *2*, 36–44. [CrossRef]

40. Kasmar, J.V. The development of a usable lexicon of environmental descriptors. *Environ. Behav.* **1970**, *2*, 153–169.

41. Mehrabian, A.; Russell, J.A. *An Approach to Environmental Psychology*; MIT Press: Cambridge, MA, USA, 1974.

42. Szokolay, S.V. *Introduction to Architectural Science: The Basis of Sustainable Design*, 2nd ed.; Elsevier: Oxford, UK, 2008.

43. Repko, A.F. *Interdisciplinary Research: Process and Theory*, 2nd ed.; SAGE: Los Angeles, CA, USA, 2012.

44. Salter, L.; Hearn, A. *Outside the Lines: Issues in Interdisciplinary Research*; McGill-Queen's University Press: Montreal, QC, Canada; Kingston, ON, Canada, 1997.

45. Nowotny, H.; Scott, P.; Gibbons, M. Introduction: 'mode 2' revisited: The new production of knowledge. *Minerva* **2003**, *41*, 179–194. [CrossRef]

46. National Research Council. *Convergence: Facilitating Transdisciplinary Integration of Life Sciences, Physical Sciences, Engineering, and Beyond*; The National Academies Press: Washington, DC, USA, 2014.

47. Hyde, R. Convergent methodologies for architectural science. *Archit. Sci. Rev.* **2013**, *56*, 101–102. [CrossRef]

48. Teddlie, C.; Tashakkori, A. *Foundations of Mixed Methods Research*; SAGE Publications Inc.: Thousand Oaks, CA, USA, 2009.

49. Groat, L.N.; Wang, D. *Architectural Research Methods*, 2nd ed.; John Wiley & Sons: New York, NY, USA, 2013.

50. Greene, J.C. Is mixed methods social inquiry a distinctive methodology? *J. Mixed Methods Res.* **2008**, *2*, 7–22. [CrossRef]

51. Fox, W. *A Theory of General Ethics: Human Relationships, Nature, and the Built Environment*; The MIT Press: Cambridge, MA, USA, 2006.

52. Cohen, R.; Standeven, M.; Bordass, B.; Leaman, A. Assessing building performance in use 1: The probe process. *Build. Res. Inf.* **2001**, *29*, 85–102. [CrossRef]

53. Williamson, T.; Soebarto, V.; Radford, A. Comfort and energy use in five australian award-winning houses: Regulated,measured and perceived. *Build. Res. Inf.* **2010**, *38*, 509–529. [CrossRef]

54. Bryman, A. The debate about quantitative and qualitative research: A question of method or epistemology? *Br. J. Sociol.* **1984**, *35*, 75–92. [CrossRef]

55. Creswell, J.W. *Research Design: Qualitative, Quantitative, and Mixed Methods Approaches*, 2nd ed.; Sage Publications: Thousand Oaks, CA, USA, 2003.

56. Teddlie, C.; Tashakkori, A. Major issues and controversies in the use of mixed methods in the social and behavioral sciences. In *Handbook of Mixed Methods in Social & Behavioral Research*; Tashakkori, A., Teddlie, C., Eds.; Sage Publishers Inc.: Thousand Oaks, CA, USA, 2003; pp. 3–50.

57. Talisse, R.B.; Aikin, S.F. *Pragmatism: A Guide for the Perplexed*; Continnum: London, UK, 2008.

58. Metcalfe, M. Pragmatic inquiry. *J. Oper. Res. Soc.* **2008**, *59*, 1091–1099. [CrossRef]

59. Baert, P. Towards a pragmatist-inspired philosophy of social science. *Acta Sociol.* **2005**, *48*, 191–203. [CrossRef]

60. Flyvbjerg, B. Phronetic planning research: Theoretical and methodological reflections. *Plan. Theory Pract.* **2004**, *5*, 283–306. [CrossRef]

61. Melles, G. An enlarged pragmatist inquiry paradigm for methodological pluralism in academic design research. *Artifact* **2008**, *2*, 3–11. [CrossRef]

62. Roussac, A.C.; de Dear, R.; Hyde, R. Quantifying the 'human factor' in office building energy efficiency: A mixed-method approach. *Archit. Sci. Rev.* **2011**, *54*, 124–131. [CrossRef]

63. Gunay, H.B.; O'Brien, W.; Beausoleil-Morrison, I. A critical review of observation studies, modeling, and simulation of adaptive occupant behaviors in offices. *Build. Environ.* **2013**, *70*, 31–47. [CrossRef]

64. Yun, H.; Nam, I.; Kim, J.; Yang, J.; Lee, K.; Sohn, J. A field study of thermal comfort for kindergarten children in korea: An assessment of existing models and preferences of children. *Build. Environ.* **2014**, *75*, 182–189. [CrossRef]

65. Downton, P. *Design Research*; RMIT University Press: Melbourne, Australia, 2003.

66. Lawson, B. *How Designers Think: The Design Process Demystified*, 4th ed.; Architectural Press: Oxford, UK, 2005.

67. Sanoff, H. *Methods of Architectural Programming*; Dowden, Hutchinson & Ross Inc.: Stroudsburg, PA, USA, 1977.

68. Hillier, B.; Leaman, A. The man-environment paradigm and its paradoxes. *Archit. Des.* **1973**, *78*, 507–511.

69. Giddens, A. *The Constitution of Society: Outline of the Theory of Structuration*; Polity Press: Cambridge, UK, 1984.

70. Hillier, B. Space and spatiality: What the built environment needs from social theory. *Build. Res. Inf.* **2008**, *36*, 216–230. [CrossRef]

71. Awan, N.; Schneider, T.; Till, J. *Spatial Agency: Other Ways of Doing Architecture*; Routledge: Abingdon, UK, 2011.

72. Latour, B. *We Have Never Been Modern*; Harvester Wheatsheaf: Hemel Hempstead, UK, 1991.

73. Stengers, I. *Thinking with Whitehead*; Harvard University Press: Cambridge, MA, USA, 2011.

74. Vischer, J.C. Towards a user-centred theory of the built environment. *Build. Res. Inf.* **2008**, *36*, 231–240. [CrossRef]

75. Erzberger, C.; Kelle, U. Making inferences in mixed methods: The rules of integration. In *Handbook of Mixed Methods in Social & Behavioral Research*; Tashakkori, A., Teddlie, C., Eds.; Sage Publishers Inc.: Thousand Oaks, CA, USA, 2003; pp. 457–488.

76. Crouch, C.; Pearce, J. *Doing Research in Design*; Berg: New York, NY, USA, 2012.

77. Brooker, G.; Weinthal, L. *The Handbook of Interior Architecture and Design*; Bloomsbury Academic: London, UK, 2013.

78. Weinthal, L. *Toward a New Interior: An Anthology of Interior Design Theory*; Princeton Architectural Press: New York, NY, USA, 2011.

79. Clemons, S.A.; Eckman, M.J. Exploring theories identified in the journal of interior design. *J. Inter. Des.* **2011**, *36*, 31–49. [CrossRef]

80. Power, J. Australian palawa buildings: Rethinking interiors and their representation. *J. Inter. Des.* **2016**, *41*, 15–32. [CrossRef]

81. Wang, Z.; Pukszta, M. Patient needs and environments for cancer infusion treatment. *J. Inter. Des.* **2017**, *42*, 13–25. [CrossRef]
82. Postiglione, G. Methods of research and criticality. In *The Handbook of Interior Architecture and Design*; Brooker, G., Weinthal, L., Eds.; Bloomsbury Academic: London, UK, 2013; pp. 59–69.
83. Sparke, P. Taste and trends. In *The Handbook of Interior Architecture and Design*; Brooker, G., Weinthal, L., Eds.; Bloomsbury Academic: London, UK, 2013; pp. 559–597.
84. McCarthy, C. Toward a definition of interiority. *Space Cult.* **2005**, *8*, 112–125. [CrossRef]
85. Attiwill, S. Towards an interior history. *IDEA J.* **2004**, *2004*, 1–8.
86. Taylor, M. *Interior Design and Architecture: Critical and Primary Sources: Sensory Engagement*; Bloomsbury: London, UK, 2013; Volume 2.
87. Popov, L.S. The social production of interiority: An activity theory approach. *IDEA J.* **2010**, 90–101.
88. Verghese, G.; Smith, D. The poetic language of interior materials and color. In *The Handbook of Interior Architecture and Design*; Brooker, G., Weinthal, L., Eds.; Bloomsbury Academic: London, UK, 2013; pp. 514–527.
89. Cantwell, C. Phenomenology and the senses in interiors. In *The Handbook of Interior Architecture and Design*; Brooker, G., Weinthal, L., Eds.; Bloomsbury Academic: London, UK, 2013; pp. 544–558.
90. Denscombe, M. *The Good Research Guide for Small-Scale Social Research Projects*, 5th ed.; Open University Press, McGraw-Hill: Maidenhead, UK, 2014.
91. GBCA. University of South Australia. Available online: http://www.gbca.org.au/members/university-of-south-australia/ (accessed on 22 January 2018).
92. Griffith Winton, A. Inhabited space: Critical theories and the domestic interior. In *The Handbook of Interior Architecture and Design*; Brooker, G., Weinthal, L., Eds.; Bloomsbury Academic: London, UK, 2013.
93. Kim, J.; Candido, C.; Thomas, L.; de Dear, R. Desk ownership in the workplace: The effect of non-territorial working on employee workplace satisfaction, perceived productivity and health. *Build. Environ.* **2016**, *103*, 203–214. [CrossRef]
94. Newman, M.E.J. *Networks: An Introduction*; Oxford University Press: Oxford UK; New York, NY, USA, 2010.

© 2018 by the author. Licensee MDPI, Basel, Switzerland. This article is an open access article distributed under the terms and conditions of the Creative Commons Attribution (CC BY) license (http://creativecommons.org/licenses/by/4.0/).

MDPI
St. Alban-Anlage 66
4052 Basel
Switzerland
Tel. +41 61 683 77 34
Fax +41 61 302 89 18
www.mdpi.com

Buildings Editorial Office
E-mail: buildings@mdpi.com
www.mdpi.com/journal/buildings

www.ingramcontent.com/pod-product-compliance
Lightning Source LLC
Chambersburg PA
CBHW041138120626
46547CB00020B/3031